Topics in Current Physics

30

Topics in Current Physics

Founded by Helmut K. V. Lotsch

1 **Beam-Foil Spectroscopy**
Editor: S. Bashkin

2 **Modern Three-Hadron Physics**
Editor: A. W. Thomas

3 **Dynamics of Solids and Liquids by Neutron Scattering**
Editors: S. W. Lovesey and T. Springer

4 **Electron Spectroscopy for Surface Analysis**
Editor: H. Ibach

5 **Structure and Collisions of Ions and Atoms**
Editor: I. A. Sellin

6 **Neutron Diffraction**
Editor: H. Dachs

7 **Monte Carlo Methods** in Statistical Physics
Editor: K. Binder

8 **Ocean Acoustics**
Editor: J. A. DeSanto

9 **Inverse Source Problems** in Optics
Editor: H. P. Baltes

10 **Synchrotron Radiation**
Techniques and Applications
Editor: C. Kunz

11 **Raman Spectroscopy** of Gases and Liquids
Editor: A. Weber

12 **Positrons in Solids**
Editor: P. Hautojärvi

13 **Computer Processing of Electron Microscope Images**
Editor: P. W. Hawkes

14 **Excitons** Editor: K. Cho

15 Physics of **Superionic Conductors**
Editor: M. B. Salamon

16 **Aerosol Microphysics I**
Particle Interactions
Editor: W. H. Marlow

17 **Solitons**
Editors: R. Bullough, P. Caudrey

18 **Magnetic Electron Lenses**
Editor: P. W. Hawkes

19 Theory of **Chemisorption**
Editor: J. R. Smith

20 **Inverse Scattering Problems** in Optics
Editor: H. P. Baltes

21 **Coherent Nonlinear Optics**
Recent Advances
Editors: M. S. Feld and V. S. Letokhov

22 **Electromagnetic Theory of Gratings**
Editor: R. Petit

23 **Structural Phase Transitions I**
Editors: K. A. Müller and H. Thomas

24 **Amorphous Solids**
Low-Temperature Properties
Editor: W. A. Phillips

25 **Mössbauer Spectroscopy II**
The Exotic Side of the Method
Editor: U. Gonser

26 **Crystal Cohesion and Conformational Energies** Editor: R. M. Metzger

27 **Dissipative Systems in Quantum Optics**
Resonance Fluorescence, Optical Bistability, Superfluorescence
Editor: R. Bonifacio

28 **The Stratospheric Aerosol Layer**
Editor: R. C. Whitten

29 **Aerosol Microphysics II**
Chemical Physics of Microparticles
Editor: W. H. Marlow

30 **Real-Space Renormalization**
Editors: T. W. Burkhardt and J. M. J. van Leeuwen

31 **Hyperfine Interactions of Radioactive Nuclei**
Editor: J. Christiansen

32 **Superconductivity in Ternary Compounds I**
Structural, Electronic, and Lattice Properties
Editors: Ø. Fischer and M. B. Maple

33 **Molecular Collision Dynamics**
Editor: J. M. Bowman

34 **Superconductivity in Ternary Compounds II**
Editors: M. B. Maple and Ø. Fischer

Real-Space Renormalization

Edited by
T. W. Burkhardt and J. M. J. van Leeuwen

With Contributions by
T. W. Burkhardt F. Family R. Jullien
J. M. J. van Leeuwen G. F. Mazenko K. A. Penson
P. Pfeuty S. Redner P. J. Reynolds M. Schick
H. E. Stanley R. H. Swendsen O. T. Valls

With 60 Figures

Springer-Verlag Berlin Heidelberg New York 1982

Dr. Theodore W. Burkhardt

Institut Laue-Langevin, 156 X, F-38042 Grenoble Cédex, France, and
Department of Physics, Temple University, Philadelphia, PA 19122, USA

Professor Dr. J. M. J. van Leeuwen

Technische Hogeschool Delft, Laboratorium voor Technische Natuurkunde, 2600 6A Delft, The Netherlands

ISBN 3-540-11459-9 Springer-Verlag Berlin Heidelberg New York
ISBN 0-387-11459-9 Springer-Verlag New York Heidelberg Berlin

Library of Congress Cataloging in Publication Data. Main entry under title: Real-space renormalization. (Topics in current physics ; 30). Includes bibliographies and index. 1. Renormalization (Physics) I. Burkhardt, T. W. (Theodore W.), 1940– . II. Leeuwen, J. M. J. van. III. Series. QC20.7.R43R43 530.4 82-5763 AACR2

Offset printing and bookbinding: Konrad Triltsch, Graphischer Betrieb, Würzburg.
2153/3130-543210

Preface

The renormalization-group approach is largely responsible for the considerable success which has been achieved in the last ten years in developing a complete quantitative theory of phase transitions. Before, there was a useful physical picture of phase transitions, but a general method for making accurate quantitative predictions was lacking. Existent theories, such as the mean-field theory of Landau, sometimes reproduce phase diagrams reliably but were known to fail qualitatively near critical points, where the critical behavior is particularly interesting because of its universal character.

In the mid 1960's Widom found that the singularities in thermodynamic quantities were well described by homogeneous functions. Kadanoff extended the homogeneity hypothesis to correlation functions and linked it to the idea of scale invariance. In the early 1970's Wilson showed how Kadanoff's rescaling could be explicitly carried out near the fixed point of a flow in Hamiltonian space. He made the first practical renormalization-group calculation of the flow induced by the elimination of short-wave-length Fourier components of the order-parameter field. The universality of the critical behavior emerges in a natural way in this approach, with a different fixed point for each universality class. The discovery by Wilson and Fisher of a systematic expansion procedure in ε for a system in $d = 4 - \varepsilon$ dimensions was followed by a cascade of calculations of critical quantities as a function of d and of the order-parameter dimensionality n. Using field-theoretic techniques, Brézin, Le Guillou, Zinn-Justin, Nickel, and others have turned the ε expansion into an astonishingly accurate tool for computing the critical behavior of three-dimensional systems.

Another class of renormalization procedures has been developed, in which the renormalization is carried out in real space rather than Fourier space. In the real-space renormalization of a system of spins on a lattice, lengths are rescaled by replacing cells of spins by single spins, following the intuitive scaling picture of Kadanoff rather closely. Unlike the Fourier-space methods, real-space renormalization works directly with a microscopic Hamiltonian and attempts to calculate non-universal information, for example phase diagrams, in addition to universal quantities. It is not a priori clear that renormalization will work on a microscopic level. There is no unique prescription for renormalizing in real space, and a considerable number of approximate renormalization techniques have been developed. There have been applications to a wide variety of different physical systems. The domain of applicability is by no means restricted to critical phenomena. Striking successes have been achieved, but there are also some general problems with the approach. This volume is a review of the present state of the art.

The literature on real-space renormalization is already too large to be reviewed in a single volume. We have included contributions on five or six general areas where, in our opinion, particularly significant developments have taken place. We decided not to repeat at length material already covered in the 1976 review article by Niemeijer and van Leeuwen in Vol.6 of the Domb-Green series *Phase Transitions and Critical Phenomena*.

In the introductory chapter we give an overview of real-space renormalization and discuss some of the fundamental problems. Griffiths-Pearce singularities, exact differential renormalization, and the phenomenological renormalization related to finite-size scaling are also considered. Chapter 2, by Burkhardt, is devoted to bond-moving and variational approximation techniques. In Chapter 3 Swendsen reviews Monte Carlo renormalization methods. Real-space renormalization procedures have been developed for quantum systems and for dynamic critical phenomena. These two topics are discussed by Pfeuty, Jullien, and Penson and by Mazenko and Valls in Chapters 4 and 5, respectively. Chapters 6 and 7 review two specific areas of application where real-space renormalization is making a significant contribution. In Chapter 6, by Schick, two-dimensional adsorbed systems are considered, and in Chapter 7, by Stanley, Reynolds, Redner, and Family, applications to percolation and polymers.

The suggestion for a Topics volume on real-space renormalization originally came from K. Binder, whom we thank for his encouragement. We also thank the contributors and H. Lotsch of Springer-Verlag for their cooperation in preparing a review of this rapidly developing field.

Delft and Philadelphia,
March 1982

T.W. Burkhardt
J.M.J. van Leeuwen

Contents

List of Contributors

Burkhardt, Theodore W.
Institut Laue-Langevin, 156 X, F-38042 Grenoble Cêdex, France, and
Department of Physics, Temple University, Philadelphia, PA 19122, USA

Family, Fereydoon
Center for Polymer Studies and Department of Physics,
Boston University, Boston, MA 02215, USA, and
Department of Physics, Emory University, Atlanta, GA 30322, USA

Jullien, Rémi
Laboratoire de Physique des Solides, Université de Paris-Sud,
Centre d'Orsay, F-91405 Orsay, France

Leeuwen van, J.M.J.
Technische Hogeschool Delft, Laboratorium voor Technische Natuurkunde,
Lorentzweg 1, 2600 GA Delft, The Netherlands

Mazenko, Gene F.
The James Franck Institute, University of Chicago, Chicago, IL 60637, USA

Penson, Karol A.
Institut für Theoretische Physik, Freie Universität Berlin,
Arnimallee 3, D-1000 Berlin 33, Fed. Rep. of Germany

Pfeuty, Pierre
Laboratoire de Physique des Solides, Université de Paris-Sud,
Centre d'Orsay, F-91405 Orsay, France

Redner, Sidney
Center for Polymer Studies and Department of Physics,
Boston University, Boston, MA 02215, USA

Reynolds, Peter J.
Center for Polymer Studies and Department of Physics,
Boston University, Boston, MA 02215, USA, and
MMRD, Lawrence Berkely Lab., University of California, Berkeley, CA 94720, USA

Schick, Michael
Department of Physics, University of Washington, Seattle, WA 98195, USA

Stanley, H. Eugene
Center for Polymer Studies and Department of Physics,
Boston University, Boston, MA 02215, USA

Swendsen, Robert H.
I.B.M. Research Laboratory, Säumerstraße 4, CH-8803 Rüschlikon, Switzerland

Valls, Oriol T.
School of Physics and Astronomy, University of Minnesota, Minneapolis, MN 55455, USA

1. Progress and Problems in Real-Space Renormalization

T. W. Burkhardt and J. M. J. van Leeuwen

With 5 Figures

The real-space renormalization approach to statistical mechanics is reviewed. A brief account of progress in the field is given. Some fundamental difficulties with the approach are discussed, in particular the proliferation of interactions, the ambiguity in the choice of the weight function, and the peculiarities in the thermodynamic limit noted by GRIFFITHS and PEARCE. Two methods which avoid the problem of proliferation of interactions are considered in some detail: differential real-space renormalization and the phenomenological renormalization related to finite-size scaling.

1.1 Introduction

Following WILSON's first application of renormalization-group ideas to critical phenomena, there has been an enormous activity in the field [1.1-9]. Renormalization techniques in both momentum and position space have been developed. In the momentum-space or field-theoretical approach, critical exponents and other universal quantities have been calculated in expansions in $\varepsilon = d - d^*$, where d is the dimension of interest and d^* is the upper or lower critical dimension. The ε expansion has lead to extremely precise estimates [1.10] of the critical exponents of several three-dimensional systems. Momentum-space renormalization-group techniques have also been applied to dynamic critical phenomena [1.11]. The starting point in a momentum-space renormalization-group calculation is a coarse-grained Hamiltonian in which details irrelevant to the universal critical behavior have been eliminated. This Hamiltonian is a functional of the order-parameter field. Renormalization results from an elimination of the short-wavelength Fourier components of the field and subsequent rescaling of lengths and of the field.

Position or real-space renormalization deals directly with the microscopic Hamiltonian. Most of the applications have considered spins on a lattice. In the renormalization, new spin variables which replace blocks of the original spins are defined and their effective interactions determined. Except in a few special applications the real-space renormalization procedures devised thus far are not exact but rely on approximations lacking an expansion parameter. So far the approach has been less successful in predicting the critical exponents of three-dimensional systems than the extrapolation procedures incorporating exact results of the ε expansion. However, for systems, particularly those with discrete spins, in low dimensions where the critical fluctuations are strong and the ε expansion not very useful, real-space renormalization is one of the most powerful of the calculational techniques of comparable simplicity currently available. As the real-space renormalization procedures are directly applicable to microscopic models of spins on a lattice with no prior coarse graining, one can calculate nonuniversal as well as universal quantities, for example, the phase diagram as a function of the interaction parameters, or the specific heat as a function of the temperature. Real-space renormalization also provides a natural description of a first-order transition [1.12, 13]. The approach has an intuitive simplicity and can often be conveniently tailored to a particular application.

This book is primarily devoted to developments in real-space renormalization which have taken place since the preparation of the 1976 review article [1.13] in the Domb-Green series. The reader seeking a detailed introduction to the subject is referred to [1.13]. Section 1.2 of this chapter contains a brief review of the real-space renormalization approach. Sections 1.3,4 summarize some of the more significant recent developments in the field not covered in [1.13]. Certain of these developments are treated in detail in other contributions to this volume. In Sect.1.5 some of the fundamental inherent difficulties in real-space renormalization are discussed, in particular the proliferation of interactions which results from the renormalization operation, the problem of choosing a suitable weight function, and the peculiarities in the thermodynamic limit pointed out by GRIFFITHS and PEARCE [1.14-16]. Sections [1.6,7] describe two successful procedures that manage to avoid these problems. In Sect.1.6 exact differential renormalization transformations [1.17-22] are considered. An exact differential transformation has been devised for the two-dimensional Ising model with nearest-neighbor interactions on a triangular lattice [1.17]. No additional couplings are generated by the renormalization, but the couplings become spatially dependent. Section 1.7 contains a discussion of "phenomenological renormalization" [1.23], a type of finite-size scaling analysis that bears some resemblance to real-space renormalization. No proliferation of interactions occurs in phenomenological renormalization either. The approximation can be systematically improved and the convergence of the results observed. Section 1.8, which completes the chapter, contains concluding remarks.

1.2 Review of Real-Space Renormalization

Real-space renormalization methods implement an intuitive picture, proposed by KADANOFF [1.24] several years before the renormalization-group breakthrough, to explain certain universal aspects of critical behavior. In this picture, cells of spins in a nearly critical system behave like individual spins in a system somewhat farther from criticality. A possible assignment of spins into cells is shown in Fig.1.1. The dots represent the original or site spins, which we denote by {s} and the crosses the cell spins {s'}. The central thermodynamic quantity of interest is the free energy

$$f = \lim_{N\to\infty} (1/N) \ln Z \quad , \tag{1.1}$$

where N is the number of s spins and the partition function Z is defined by

$$Z = \sum_{s} \exp H(s) \quad . \tag{1.2}$$

H(s) is the dimensionless Hamiltonian (in which a factor $-(k_BT)^{-1}$ has been incorporated) of the system of s spins. The sum in (1.2) is over all configurations of the s spins.

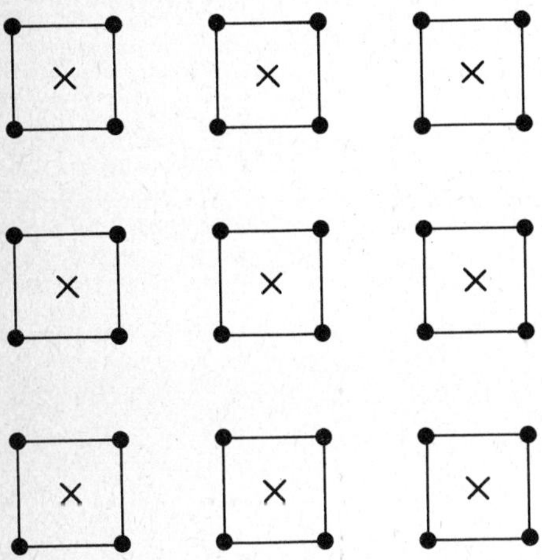

Fig.1.1. Possible assignment of spins to cells. The dots represent site spins and the crosses cell spins. There are four sites per cell

Adding an arbitrary spin-independent constant to H(s) changes the free energy by an additive constant. It is convenient to choose a convention such as $\sum_s H(s) = 0$ or H = 0 for a particular spin configuration in order to define the Hamiltonian H(s) uniquely.

In real-space renormalization the interactions between the cell spins are calculated from the interactions between the site spins. A Hamiltonian G + H'(s') for the cell spins (where G is independent of the spin configurations and the zeroes of H'(s') and H(s) are fixed according one of the conventions mentioned above) is defined by the renormalization transformation

$$\exp[G + H'(s')] = \sum_{s} P(s',s) \exp H(s) \quad . \tag{1.3}$$

The weight function P(s',s), which couples the cell and site spins, satisfies

$$\sum_{s'} P(s',s) = 1 \quad , \tag{1.4}$$

which insures that the partition functions of the cell and site-spin systems are the same. The condition (1.4) is not very restrictive, and a variety of weight functions [1.13] have been tried. Generally (some exceptions are noted below), the transformation is designed so that the s' and s spins are spins of the same type, for example, in applications to the Ising model both the s' and s spins take the values ±1. Evaluating the sum over configurations in (1.3) reduces the number of spins and enlarges the lattice constant. In momentum-space renormalization [1.1-9] degrees of freedom are eliminated and the length scale expanded by integrating the Bolztmann factor exp H of the coarse-grained Hamiltonian over short-wavelength fluctuations of the spin variables.

It is useful to expand H(s) in the form $H(s) = \sum_{\alpha} K_{\alpha} \Omega_{\alpha}(s)$, where the K_{α} are coupling constants and the $\Omega_{\alpha}(s)$ a complete set of interactions generated by repeated application of (1.3) from the initial Hamiltonian of interest. Because of the proliferation of interactions under the renormalization operation, the set generally contains an infinite number of elements, but in approximate calculations only a finite subset is taken into account. H' can also be expanded in terms of the Ω_{α}, with coupling constants K'_{α}. Equation (1. 3) implies a set of equations

$$\underline{K}' = \underline{K}'(\underline{K}) \tag{1.5}$$

relating the coupling constants of the original and renormalized spins. Here $\underline{K}$ and $\underline{K}'$ denote the entire sets of K_{α} and K'_{α}. The transformation law

$$f(\underline{K}) = g(\underline{K}) + \frac{1}{b^d} f(\underline{K}') \tag{1.6}$$

for the free energy per spin also follows from (1.3,4). In deriving (1.6), one has written $g = G/N$ and $b^d = N/N'$, b being the length rescaling factor, and the thermodynamic limit is implicitly taken to be $N \to \infty$. From (1.6) one sees that $g(\underline{K})$ is the contribution to the free energy from the degrees of freedom eliminated in the renormalization step.

Once the functions $\underline{K}'(\underline{K})$ and $g(\underline{K})$ are known, the free energy may be calculated with the formula

$$f(K) = \sum_{n=0}^{\infty} \frac{1}{b^{nd}} g(\underline{K}^{(n)}) + \lim_{m\to\infty} \frac{1}{b^{md}} f(\underline{K}^{(m)}) \quad , \tag{1.7}$$

obtained by iterating (1.5,6). $\underline{K}^{(n)}$ in (1.7) denotes the set of coupling constants after n iterations. In the renormalization-group approach [1.1-9,13] one can calculate critical exponents separately from the free energy. The critical surface of continuous transitions belonging to a particular universality class is assumed to map onto a fixed point $\underline{K}^*$ of (1.5). The scaling indices y_μ that determine critical exponents and the corresponding relevant variables are determined from the leading eigenvalues b^{y_μ} and eigenvectors of the matrix

$$T_{\alpha\beta} = \partial K'_\alpha/\partial K_\beta \Big|_{\underline{K}=\underline{K}^*} \quad . \tag{1.8}$$

A fixed point corresponding to a first-order, rather than a continuous, transition generally has one or more eigenvalues $y_\mu = d$ consistent with discontinuities in the derivatives of the free energy [1.12,13]. For a more extensive discussion of all of these points the reader is referred to [1.13] and references therein.

In general the sum over configurations in the renormalization transformation (1.3) is no more tractable than the sum in the partition function (1.2). However, advocates of the renormalization group regard (1.3) as a better starting point for approximations. The partition function (1.2) is a nonanalytic function of the coupling constants in H(s). The critical exponents that characterize these nonanalyticities are very sensitive to approximations that render the sum over configurations tractable. For example, conventional approximations such as the cumulant, quasi-chemical, and Kikuchi cluster-variation approximations [1.25] invariably lead to mean-field critical exponents. In the sum in (1.3) only the subset of s configurations consistent with a given configuration of the s' spins is effectively included. Until recently the weight function was assumed to eliminate all "dangerous" long-wavelength critical fluctuations from the sum in (1.3), making the right-hand side a smooth nonsingular function of the coupling constants much less sensitive to approximations in evaluating the sum over configurations. The success of real-space renormalization-group calculations involving rather drastic approximations lends support to this point of view.

Recently GRIFFITHS and PEARCE [1.14-16] pointed out that for most of the weight functions that have been used so far, there are configurations of the s' spins that do not eliminate critical fluctuations of the s spins, contrary to what was generally accepted. The assumption that $g(\underline{K})$ and the $\underline{K}'$ are globally nonsingular functions of the $\underline{K}$ appears to be incorrect in the case of exact calculations in the thermodynamic limit. However, no compelling evidence for singularities of the renormalization transformation in the critical region of the system of interest has yet been presented. We discuss the origin and significance of the peculiarities in the thermodynamic limit noted by GRIFFITHS and PEARCE in more detail in Sect.1.5.

The first attempts to use real-space renormalization to calculate critical properties are reviewed in [1.13]. In that work renormalization transformations based on cumulant and cluster approximations were applied to the two-dimensional Ising model. In the cumulant approximations the intracell interactions are treated exactly and the intercell interactions to low order in perturbation theory. In cluster approximations the renormalization transformation is worked out exactly for finite subsets rather than the infinite lattice. Both linear weight functions, such as in ordinary decimation, and nonlinear weight functions, such as the majority rule (in which the cell spin is the same as the majority of the site spins in the cell) have been considered. The cumulant expansion appears to converge rather slowly if at all. For large clusters the cluster approximation appears to work quite well. In both of these approximations only a limited number of short-range interactions must be taken into account. The number increases as the approximations are improved by considering higher orders in perturbation theory or larger clusters.

1.3 New Renormalization Methods

In this section we summarize some of the more significant developments in real-space renormalization since the review article [1.13] was prepared. Several more approximation methods have been proposed, and real-space renormalization techniques have been applied to a much wider variety of problems. We begin with a discussion of new approximation methods.

1.3.1 Bond-Moving and Variational Methods

A considerable amount of work on bond-moving and variational methods in real-space renormalization, largely based on ideas of KADANOFF and MIGDAL, has been carried out since 1975. This field is reviewed by BURKHARDT in Chap.2 of this volume. KADANOFF [1.26,27] first indicated how variational renormalization transformations that give upper and lower bounds to the exact free energy can be devised. Transformations in which the only approximation is "bond moving" overestimate the partition function and thus yield an upper bound to the free energy (1.1) (a lower bound if a minus sign is included in the definition of the free energy). KADANOFF invented a simple bond-moving transformation with a variational parameter in the weight function that predicts the critical exponents of a considerable number of models with astonishing accuracy when the variational parameter is fixed at the value that optimizes the free energy at the appropriate fixed point. The success of this transformation is marred by an inconsistency in the application of the variational criterion. Allowing the variational parameter to change in each application of the renormalization transformation improves the bound on the free energy but worsens the critical exponents and leads to other fundamental difficulties [1.28,29]. Why KADANOFF's original procedure for calculating critical exponents works so well is not well understood.

MIGDAL [1.30] has invented an approximate renormalization transformation which is unequalled in simplicity and has been widely used. It can readily be applied in arbitrary dimension to systems with continuous as well as discrete spins. The transformation works especially well just above the lower critical dimensionality, where the critical temperature is almost zero. The approximation preserves the dual symmetries of certain two-dimensional discrete-spin models and thus yields the phase diagram exactly, although the corresponding predictions for critical exponents are not always very accurate. KADANOFF [1.31] has rederived MIGDAL's transformation in terms of bond-moving and decimation operations.

1.3.2 Monte Carlo Renormalization

The idea of combining Monte Carlo simulations and real-space renormalization was suggested by MA [1.32] and subsequently developed in several different forms [1.33-36]. A simulation of the spins in a large system can also be interpreted in terms of block-spin variables interacting according to a renormalized Hamiltonian. If the system is sufficiently large, a single simulation contains information about several stages of the renormalization process. In a method due to SWENDSEN [1.35], one extracts the matrix $K_\alpha^{(n)}/K_\beta^{(n-1)}$ for several renormalization steps from correlation functions determined in a simulation with the initial Hamiltonian, and one looks for convergence. In principle, the simulation can be analyzed using as many coupling constants as desired. Only a limited number of renormalization steps can be taken before the cell size becomes comparable with the size of the system. However, excellent quantitative results have been obtained despite this restriction. A chapter by SWENDSEN on Monte Carlo renormalization has also been included in this volume.

1.3.3 Exact Differential Transformations

Exact differential renormalization transformations have been devised for the Ising model in one and two dimensions [1.17-19] and for the Gaussian model in d dimensions [1.20,21]. In its present form the approach is limited to system with nearest-neighbor interactions for which a star-triangle transformation can be performed, e.g., the Ising model on a triangular lattice. An unusual feature is the spatial dependence of the coupling constants induced in initially homogeneous systems by the renormalization. For all of the models listed above the fixed point of the transformation can be calculated analytically and the thermal scaling index $y_t = \nu^{-1}$ determined exactly. Section 1.6 is devoted to exact differential renormalization methods.

1.3.4 Phenomenological Renormalization

This method, invented by NIGHTINGALE [1.23], is a type of finite-size scaling analysis that bears some resemblance to real-space renormalization. One obtains predictions of critical couplings and exponents from a scaling relation between the correlation lengths in systems with different finite sizes. The method has mainly been applied to two-dimensional strips with infinite length and finite width, using the transfer matrix to calculate the correlation length. The approximation, which can be systematically improved by increasing the system size, gives very reliable quantitative results. Phenomenological renormalization is discussed in more detail in Sect.1.7.

1.4 New Applications

We will not attempt to review thoroughly the vast literature on applications of real-space renormalization. References to some of the applications may be found elsewhere in this book. Here we mention several broad fields where the approach is playing a significant role. The domain of applicability is by no means restricted to static critical phenomena.

1.4.1 Adsorbed Systems

The chapter by SCHICK in this volume is concerned with applications to phase transitions in adsorbed systems, a field where there has been a fruitful cooperation between theory and experiment. A number of models that have considerable theoretical interest, for example, the q-state Potts model with $q = 2,3,4$ and models for the commensurate-incommensurate transition, may be realized in adsorbed systems. Real-space renormalization has been useful in analyzing experimental data. SCHICK's article also discusses the picture of the crossover from a second-order to a first-order transition in the Potts model that emerges from the renormalization-group analysis of NIENHUIS et al. [1.37-39], in which vacancies are included. This picture, which solves a long-standing puzzle, relates critical and tricritical exponents, and has rather general significance, is also considered in Sect.1.5.

1.4.2 Applications to Quantum Systems

The chapter by PFEUTY et al. in this volume reviews applications of real-space renormalization to quantum systems. Renormalization techniques have been used to study both finite-temperature and ground-state properties of quantum spin systems, particularly "phase transitions" in the ground state that take place as interaction constants in the Hamiltonian are varied. The well-known correspondence [1.40,41] between ground-state transitions in certain d-dimensional quantum systems and finite-temperature transitions in related d+1-dimensional classical systems suggests the usefulness of renormalization techniques in ground-state calculations. In a widely used method for calculating ground-state properties, a system of quantum spins is divided into cells. Each cell is replaced by a single quantum spin, with the renormalized interactions chosen so that the lowest energy levels of the site and cell-spin systems coincide. WILSON's study of the Kondo problem [1.4] involves a renormalization operation which is somewhat similar in spirit.

1.4.3 Percolation and Polymers

Applications of real-space renormalization to percolation and polymers are considered in the chapter in this book by STANLEY et al. The equivalence between percolation [1.42] and the q-state Potts model as $q \to 1$ and the self-avoiding walk [1.43] and the n-vector model as $n \to 0$ suggests the usefulness of the renormalization-group approach. Most of the work in this field does not explicit these equivalences but involves forms of real-space rescaling immediately related to the physical problem of interest.

1.4.4 Dynamic Real-Space Renormalization

This topic is considered in the chapter by MAZENKO and VALLS. There have been several attempts to use real-space renormalization to study the critical dynamics of the kinetic Ising model, but the basic procedure to be followed is not as apparent as in static renormalization (see, for example, [1.44]).

1.4.5 The Kosterlitz-Thouless Transition

A substantial amount of work [1.45-48] has been devoted to the peculiar transition exhibited by planar rotators on a two-dimensional lattice. This model is related to a host of other two-dimensional models [1.45-50]. KOSTERLITZ [1.46] has analyzed the transition using differential real-space renormalization equations for the two-dimensional Coulomb-gas system.

1.4.6 Field-Theoretical Applications

Finally we note that real-space renormalization methods have also been applied to certain field-theoretical models, for example, lattice gauge theories of quarks and strings. The lattice introduces a cutoff which prevents divergences. For an introduction to field-theoretical applications the reader is referred to [1.30,51-53].

1.5 Fundamental Problems

In the field-theoretical approach to critical phenomena [1.1-9], systematic expansions in ε, the deviation of the dimension from the upper or lower critical dimension, have been developed. Real-space renormalization-group methods have most frequently been applied to intermediate-dimensional systems, where critical fluctuations are strong. There is no obvious expansion parameter, and except in a few special exactly tractable cases, the calculational schemes involve drastic truncations. We have already mentioned the problem of infinite proliferation of interactions, which is generally encountered, and reviewed some of the approximation methods that have been devised.

In the absence of a small expansion parameter the best way to proceed is far from obvious. The mild restriction (1.4) allows one a great deal of freedom in choosing the weight function. Recent work by NIENHUIS et al. [1.37-39] indicates that the choice of the weight function is a considerably more subtle matter than was widely suspected. To explain the crossover from a second-order to a first-order transition in the Potts model as the number of states increases, they found it necessary to enlarge the space of lattice variables on renormalization to include vacancies. GRIFFITHS and PEARCE [1.14-16] have pointed out that with most of the weight functions used so far, the assumption that the renormalization transformation only involves nonsingular functions of the coupling constants is not globally valid in the thermodynamic limit.

Before turning in more detail to the problem of choosing the weight function and the singularities pointed out by GRIFFITHS and PEARCE, we note several general limitations of the real-space renormalization-group approach in its present form.

A serious problem is the absence of error estimation procedures within the framework of real-space renormalization. It is straightforward in principle but extremely difficult in practice to systematically improve most of the approximate transformations sufficiently to draw conclusions about the convergence. Thus, the quantitative accuracy of the predictions is uncertain. In the case of phenomenological renormalization, a type of finite-size scaling analysis discussed in Sect. 1.7, this problem is less acute. The convergence of the predictions can be studied as the finite size is increased.

The approximation methods used to study finite-temperature critical phenomena in low-dimensional discrete-spin systems with short-range forces are also applicable in principle to systems with more complicated spin variables, for example, continuous variables or quantum spins. In practice, the mathematical manipulations become much more formidable, and only limited progress has been made. Some applications to systems with continuous spin variables are considered in [1.27,30,36,54]. References to work on finite-temperature critical phenomena in quantum systems are given in the chapter by PFEUTY et al. in this volume.

Practically all of the real-space renormalization approximation methods developed so far are only applicable in the case of short-range forces. Aside from exact analyses of the Ising model with constant infinite-range couplings [1.22,55], where mean-field theory is exact, there has been little progress in applying real-space renormalization to systems with long-range interactions.

In calculations based on existing approximation methods in real-space renormalization, the essential role of the dimension of the system in critical phenomena does not emerge clearly. None of the approximate transformations developed for the Ising model with short-range ferromagnetic forces gives mean-field critical exponents for $d > 4$, although the predictions of the Kadanoff variational method [1.26,27] are quite close. The truncation of interactions in the approximate transformations is so severe that no counterpart of the crossing of the Gaussian and non-Gaussian fixed points at d = 4 in the ε expansion [1.1-9] is generally found. (See, however, [1.56,36]).

Finally, we mention that difficulties have been encountered in attempting to compute spatially-dependent quantities, such as the spin-spin correlation function, with some real-space approximation methods. (A general analysis of the spin-spin correlation function within the framework of real-space renormalization is given in [1.13]). Certain cumulant and cluster approximations with nonlinear weight functions yield magnetization profiles and correlation functions which decay algebraically rather than exponentially and show other unphysical features [1.57,58]. MAZENKO et al. [1.58] have modified the first-order cumulant approximation to avoid this problem.

1.5.1 Choice of the Weight Function

At first glance one appears to have an enormous freedom in choosing the weight function in real-space renormalization. Condition (1.4) and the additional requirement $P(s',s) > 0$, which is sometimes imposed to insure that only real Hamiltonians are generated, are not very restrictive[1]. However, it should not be assumed that each weight function satisfying these two conditions leads to a well-defined transformation with a fixed point which describes the critical properties consistently. It has been pointed out [1.4,13,59] that any linear weight function (including ordinary decimation) is inconsistent with the expected scaling properties of the correlation function at the fixed point unless a coupling parameter with a value precisely dictated by the exponent η is incorporated in the weight function. Exact transformations with all the desired properties may only exist for a very restricted class of weight functions. However, apart from the work on differential renormalization reviewed in Sect.1.6, very little is known about the properties of exact real-space transformations, and in practice on must choose the weight function from a vast range of possibilities.

The only weight functions that have received much attention are local, i.e., they only couple an s' spin to one or more s spins in the immediate vicinity. There is an obvious reason for preferring local weight functions. The weight function may be viewed as a Hamiltonian coupling the systems of s and s' spins. In order to have a renormalized Hamiltonian with short-range interactions, long-range couplings in the weight function should be avoided. However, despite the locality these weight functions generally lead to infinite proliferation of interactions on renormalizing. The possibility of combatting the proliferation with more complicated weight func-

---

1 An additional condition on the weight function follows from dynamic considerations. In the chapter by MAZENKO and VALLS the requirement that the weight function be an eigenfunction of the dynamic operator is imposed.

tions, i.e., of transferring the complexity from the Hamiltonian to the weight function, has not been thoroughly explored. However, a strategy along these lines is incorporated in exact differential transformations (Sect.1.6), where the weight function is chosen so that the product P(s',s) exp[H(s)] rather than P(s',s) is simple.

The physical predictions of approximate renormalization calculations depend, of course, on the choice of the weight function. Thus it is desirable to have criteria for making an optimal choice. The importance of selecting a weight function that maps each of the ground states of the system onto itself is stressed in [1.60]. Several different criteria that further determine the weight function have been proposed. One such criterion involves choosing an adjustable parameter in the weight function so that the magnetic scaling indices determined from the scaling properties of the free energy and the correlation function are the same [1.4,13,59]. In the variational approach to real-space renormalization (Sect.1.3.1) one or more adjustable parameters in the weight function are chosen so as to optimize an upper or lower bound to the exact free energy. A third type of criterion picks adjustable parameters in the weight function so that the group structure of exact renormalization transformations, i.e., the invariance of physical quantities with respect to changes in the length rescaling factor, is preserved in a weakened form in an approximate calculation [1.61].

Renormalization transformations that leave the partition function invariant, i.e., satisfy (1.3,4), but change the nature of the spin variables have also been considered. Such transformations only make sense where no change in the universality class is anticipated. In [1.54,62,63] initial restructuring transformations that map models in the Ising universality class onto equivalent models with simpler spin-1/2 Ising variables are devised. In important recent work on the q-state Potts model [1.37-39], NIENHUIS et al. choose a transformation that enlarges the space of lattice variables. The initial Potts model is mapped onto a Potts lattice gas in which there is either a Potts spin or a vacancy at each site. All previous calculations using real-space renormalization methods, none of which included vacancies, failed to show the first-order transition for $q > 4$, $d = 2$ known to occur from exact work by BAXTER. In the picture of NIENHUIS et al. the change in the order of the transition is associated with the annihilation of critical and tricritical fixed points of the Potts lattice gas. For a more detailed account of renormalization-group studies of the Potts model including vacancies, we refer to the chapter by SCHICK in this volume.

The success of NIENHUIS et al. in predicting tricritical exponents and the solution [1.50] of a one-dimensional quantum model that corresponds to the two-dimensional Potts model [1.64] have confirmed the picture of annihilating critical and tricritical fixed points. In the quantum calculation the critical and tricritical points arise in a natural way without the inclusion of vacancies in the renormalization.

NIENHUIS et al. [1.37] and BERKER and coworkers [1.65,66] argue that the inclusion of vacancies is intuitively reasonable, that clusters of Potts spins in different states behave more like vacancies than single Potts spins, and that the condensation of these vacancies is the mechanism which turns the transition to first order as q increases. Their work on the Potts model has significant implications for real-space renormalization. Choosing a weight function to capture the essential physics in a simple approximate transformation appears to be a rather subtle matter demanding considerable physical insight. In some applications [1.67] it appears necessary to introduce extra degrees of freedom per lattice site in the renormalization step.

1.5.2 Griffiths-Pearce Peculiarities

A fundamental assumption in the renormalization-group approach to critical phenomena is that the renormalized coupling constants are nonsingular functions of the original couplings, at least in the neighborhood of the fixed point of interest. In momentum-space renormalization this assumption is justified by arguing that only the contributions of short-wavelength fluctuations, not the "dangerous" long-wavelength critical fluctuations, are summed in a single renormalization step. In real-space renormalization the separation into short- and long-wavelength fluctuations is not as precise as in momentum-space renormalization. In the renormalization procedure corresponding to (1.3), one holds the s' spins fixed and calculates a sum over all configurations of the s spins weighted with the function P(s',s), which couples the s' and s spins. The right side of (1.3) is a nonsingular function of the coupling constants in H(s) only if the coupling to the fixed s' spins eliminates critical fluctuations from the sum.

Recently GRIFFITHS and PEARCE [1.14-16] noted that with most of the weight functions that have been widely used, there are configurations of the s' spins for which the right side of (1.3) is singular. In the case of an exact calculation in the thermodynamic limit, the assumption that the renormalization transformation only involves nonsingular functions of the coupling constants does not appear to be globally valid. The singularities pointed out by GRIFFITHS and PEARCE are generally not encountered in approximate calculations but are eliminated by the decoupling procedures on which the approximation methods are based.

For a detailed discussion of the singularities in the renormalization transformation the reader is referred to [1.14-16]. Here we mention some simple situations where the origin of the singularities is particularly easy to understand.

Consider [1.15] an infinite spin system with ferromagnetic nearest-neighbor interactions K on a square lattice, as indicated in Fig.1.2a. In the renormalization step all of the s spins which are represented by black dots are removed by an ordinary decimation, leaving the s' spins, which are denoted by empty circles. In the particular case of an antiferromagnetic configuration of the s' spins, the

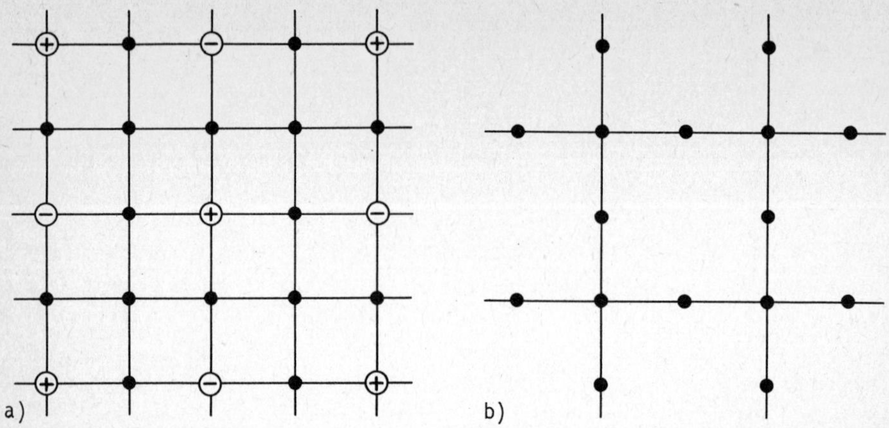

Fig.1.2a,b. In the renormalization step the s spins (black dots) in (a) are decimated. For an antiferromagnetic configuration of the s' spins (empty circles), there is no net coupling between the s and s' spins. Decimating is equivalent to evaluating the partition function of the decorated model shown in (b)

s spins no longer interact with the s' spins. The right side of (1.3) is the same as the partition function of the decorated square-lattice system shown in Fig.1.2b. The critical coupling for a phase transition in this system, which gives the location of the Griffiths-Pearce singularity, is greater than the critical coupling of the system on a square lattice with which one began.

One may argue that this example is not very relevant, since the ordinary decimation transformation is inconsistent with an algebraic decay of the spin-spin correlation function [1.4,13,59], which is the rule rather than the exception at criticality. As an example of a nonlinear weight function [1.4,13,59], for which no such inconsistency has been demonstrated, we consider

$$P(s,s') = \prod_i \frac{\exp[ps'_i(s_{i1} + .. + s_{in})]}{2\cosh[p(s_{i1} + .. + s_{in})]} \quad , \tag{1.9}$$

where s' and s are Ising spins that take the values ±1. This weight function, which assigns a cell spin s' to each of the cells of n s spins into which the system is divided, was first used by KADANOFF [1.26,27]. In the limit $p \to \infty$ (1.9) reduces to the majority-rule [1.13] weight function.

Using (1.9), one may write (1.3) in the form

$$\begin{aligned} G + H'(s') &= N'f_p(s') \\ &= \ln \sum_s \exp\left\{p\left[\sum_i s'_i(s_{i1} + .. + s_{in}) + H_{eff}(s)\right]\right\} \quad , \end{aligned} \tag{1.10}$$

where

$$H_{eff}(s) = H(s) - \sum_i \ln\{2\cosh[p(s_{i1} + .. + s_{in})]\} \quad . \tag{1.11}$$

The quantity $f_p(s')$ in (1.10) is formally identical with the free energy per spin of an Ising system with Hamiltonian $H_{eff}(s)$ subject to local fields $p_i = ps_i'$ acting on each spin in cell i. The logarithmic terms in $H_{eff}(s)$ oppose a ferromagnetic ordering. Thus for a ferromagnetically coupled H(s) and a finite p, one would expect the system with Hamiltonian H_{eff} to undergo a paramagnetic-ferromagnetic transition at a lower critical temperature (greater critical coupling) than the system with Hamiltonian H.

It can readily be seen that the right side of (1.10) exhibits Griffiths-Pearce singularities for various configurations of the s' spins. For example, if all the s' spins are up, $f_p(s')$ reduces to the free energy of a system with Hamiltonian $H_{eff}(s)$ in a uniform magnetic field p and has the same singularities. If the s' spins are ordered antiferromagnetically, $f_p(s')$ has the singularities associated with a staggered rather than a uniform field on the cells. In either case the singularities occur outside the critical region of the physical system of interest having a Hamiltonian H(s) rather than $H_{eff}(s) + p\sum_i s_i'(s_{i1} + \dots + s_{in})$.

We now consider the location of the Griffiths-Pearce singularities in the coupling constants that parametrize the Hamiltonian. An arbitrary Ising Hamiltonian can be expanded in the form

$$G + H'(s') = Ng + \sum_i K_i' s_i' + \sum_{i<j} K_{ij}' s_i' s_j' + \sum_{i<j<k} K_{ijk}' s_i' s_j' s_k' + \dots \quad . \tag{1.12}$$

The coupling constants may be obtained from the renormalized Hamiltonian using the inversion formula [1.13]

$$\begin{bmatrix} Ng \\ K_i' \\ K_{ij}' \\ \cdot\cdot \end{bmatrix} = \frac{1}{2^{N'}} \sum_{s'} \begin{bmatrix} 1 \\ s_i' \\ s_i' s_j' \\ \cdot\cdot \end{bmatrix} (G + H'(s')) \quad . \tag{1.13}$$

Assuming that H and H' only contain interactions between even numbers of spins and replacing G + H' in (1.13) by $N'f_p$, one finds

$$g = b^{-d}[f_p]_{av} \quad , \tag{1.14}$$

$$K_{ij}' = \frac{1}{2} N'[f_p(\delta_{s_i's_j'} - \delta_{s_i',-s_j'})]_{av} \quad , \tag{1.15}$$

and similar formulae for the other couplings. The square brackets $[\ ]_{av}$ indicate an average over the local fields $p_i = ps_i'$. The Kronecker δ functions impose correlations on the fields at sites i and j.

Equations (1.14,15) establish a connection between the Griffiths-Pearce singularities of the coupling constants and phase transitions in Ising models with quenched random fields explored in [1.68]. According to (1.14) the constant contribution $g(\underline{K})$ [see (1.6)] is singular where the random-field free energy defined by

(1.10,11,14) is singular. One anticipates an $O(N^{-1})$ correction to the free energy on including correlations between the quenched fields as in (1.15) but not a change in the location of the singularities. All of the interaction constants $\underline{K}'$ coupling finite numbers of spins are singular where $[f_p]_{av}$ is singular, i.e., where there is a phase transition in the quenched random-field problem.

GRIFFITHS and PEARCE [1.14-16] expand the renormalized Hamiltonian in the form

$$G + H'(s) = N\bar{g} + \sum_i \Phi'_i n'_i + \sum_{i<j} \Phi'_{ij} n'_i n'_j + \sum_{i<j<k} \Phi'_{ijk} n'_i n'_j n'_k + \dots \qquad (1.16)$$

where the $n'_i = (1 + s'_i)/2$ are lattice-gas occupation numbers which take the values 1 and 0. From (1.10) and (1.16) one sees that

$$\bar{g} = \frac{1}{N} [G + H'(s')]|_{n'=0} = b^{-d} f_p(s')|_{s'=-1} \; , \qquad (1.17)$$

i.e., the constant term on the right side of (1.16) is proportional to the free energy of the system with Hamiltonian $H_{eff}(s) - p\sum_i (s_{i1} + \dots + s_{in})$. Unlike the singularities in g and the $\underline{K}'$ discussed above, the singularities in $\bar{g}$ (and in the $\underline{\Phi}'$ coupling finite numbers of particles as well) are associated with a uniform rather than a random field.

In terms of the Ising couplings in (1.12) $\bar{g}$ has the expansion

$$\bar{g} = g - \frac{1}{N} \left(\sum_i K'_i - \sum_{i<j} K'_{ij} + \sum_{i<j<k} K'_{ijk} - \dots \right) \; . \qquad (1.18)$$

Note that infinite-spin couplings must also be included in the expansion (1.18) if the sum is to have constant-field rather than random-field singularities.

Neither the random-field nor the constant-field Griffiths-Pearce singularities discussed here appear to violate the assumption of a nonsingular renormalization transformation in the critical region of the system of interest. Adding the logarithmic terms to H(s) in (1.11) depresses the critical temperature for the paramagnetic-ferromagnetic transition. The inclusion of a random-field further lowers the critical temperature. (Below a lower critical dimensionality, which has been suggested to be 2 [1.69] or 3 [1.70], the random field is thought to reduce the critical temperature to T = 0). In the constant-field case the Griffiths-Pearce singularities occur at sufficiently low temperatures where the fictitious field associated with the weight function is cancelled by a field in H(s), in which case H(s) is not critical.

In the case of the mean-field Ising model with constant infinite-range interactions, the renormalization transformation can be worked out exactly. STELLA [1.71] has verified the existence of random and constant-field Griffiths-Pearce singularities of the sort discussed above and shown that no such singularities are encountered in the critical region.

It is possible to invent weight functions that do not lead to Griffiths-Pearce singularities. The exact differential transformations discussed in Sect.1.6 are completely free of such singularities. In these transformations one writes $\exp H_c(s',s) = P(s',s)\exp H(s)$ and in the case of the Ising model on a triangular lattice chooses $H_c(s',s) = \sum_i s_i'(p_{i1}s_{i1} + p_{i2}s_{i2} + p_{i3}s_{i3})$, where s_{i1}, s_{i2}, and s_{i3} are the s spins closest to s_i'. In $H_c(s',s)$ the s spins are coupled directly with the s' spins but not with each other. Thus for arbitrary configurations of the s' spins, the right side of (1.3) is a nonsingular function of the $p_{i\alpha}$, which in turn are nonsingular functions of the coupling constants in H(s).

Intuitively one would expect differential renormalization procedures to be less susceptible to Griffiths-Pearce singularities than the usual site-cell and decimation procedures [1.13] that reduce the number of spins by a finite fraction in each renormalization step. An infinitesimal fraction of the degrees of freedom of the system is less likely to exhibit dangerous critical fluctuations with the remaining degrees of freedom held fixed than a finite fraction.

Whether exact renormalization transformations that have fixed points and all the desirable properties one normally assumes really exist for the weight functions commonly used in finite-difference, rather than differential, renormalization is a question still completely open. From the work of GRIFFITHS and PEARCE it has become fairly clear that such transformations would not be *globally* nonsingular in the coupling constants. However, no compelling evidence for singularities in the critical region, which would cast doubts on the standard procedure for calculating critical exponents, has yet been found. Even before the work of GRIFFITHS and PEARCE there was evidence that certain weight functions that seem reasonable near a particular fixed point are not globally applicable. For example, linear weight functions consistent with the expected critical scaling properties of the spin-spin correlation function do not always satisfy $P(s',s) > 0$ for arbitrary configurations and may lead to complex coupling constants at low temperatures [1.4,13]. Singularities in the renormalization transformation may also occur where the blocking procedure fails to treat the low-temperature behavior correctly [1.13,60]. For example, if the weight function and the cells are chosen to map a ferromagnetic, but not an antiferromagnetic, ground state onto itself, the transformation is presumably singular on the paramagnetic-antiferromagnetic critical surface [1.13,60].

1.6 Exact Differential Real-Space Renormalization

Intuitively one might expect that devising an accurate renormalization scheme becomes simpler in the limit in which the change in length scale is infinitesimal rather than finite, since only an infinitesimal subset of the degrees of freedom of the system need then be summed in the renormalization step. In momentum-space renormalization both finite and infinitesimal renormalization-group procedures have been utilized [1.1-9].

In most of the approximate real-space renormalization methods developed thus far, the change in length scale is finite rather than infinitesimal. In site-cell transformations it is natural to assign an integral number of sites to a cell, and in decimation transformations to eliminate a finite fraction of the spins in the system. MIGDAL's renormalization transformation [1.30] is an example of an approximate transformation that can be set up for an arbitrary length rescaling factor b. In the limit in which b is only infinitesimally larger than 1, the evolution of the nearest-neighbor interaction is governed by a differential rather than a difference equation.

Recently HILHORST et al. [1.17] have derived exact renormalization-group equations for the two-dimensional Ising model for the first time. Their renormalization procedure is an example of differential real-space renormalization. Although their approach avoids the proliferation of interactions, it is beset by another complication. The renormalization generates spatially-dependent couplings. Nevertheless the fixed point can be located analytically and the thermal scaling index $y_t = \nu^{-1}$ determined exactly.

In the remainder of Sect.1.6 we first derive and briefly analyze the exact differential renormalization transformation of HILHORST et al. The reader is referred to the original articles [1.17] for more details. We then mention some other applications and discuss advantages and limitations of the approach.

1.6.1 The Two-Dimensional Ising Model

In considering infinitesimal real-space renormalization it is convenient to write the transformation equations (1.3,4) in a more symmetric form. Expressing the weight function in terms of the coupling Hamiltonian $H_c(s',s)$ defined by

$$\exp[H_c(s',s)] = P(s',s)\,\exp[H(s)] \quad , \tag{1.19}$$

one obtains the formulae

$$\exp[G + H'(s')] = \sum_{s} \exp[H_c(s',s)] \quad , \tag{1.20}$$

$$\exp[H(s)] = \sum_{s'} \exp[H_c(s',s)] \quad , \tag{1.21}$$

which for an arbitrary $H_c(s',s)$ leave the partition function invariant. If $H_c(s',s)$ is chosen to be symmetric in s' and s, (1.20,21) correspond to the identity transformation G = 0, H' = H. HILHORST et al. utilize a coupling Hamiltonian that is not quite symmetric and that generates an infinitesimal change in the Hamiltonian.

Following [1.17] we consider a triangular lattice in the shape of an equilateral triangle as in Fig.1.3. The dots indicate the s spins and the crosses the s' spins. The lines connecting the s and s' spins represent coupling parameters in $H_c(s',s)$, which are described in more detail below. The lengths of the sides of the triangles

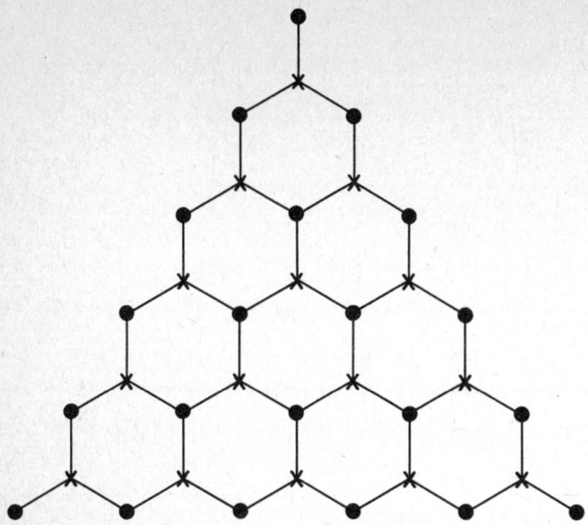

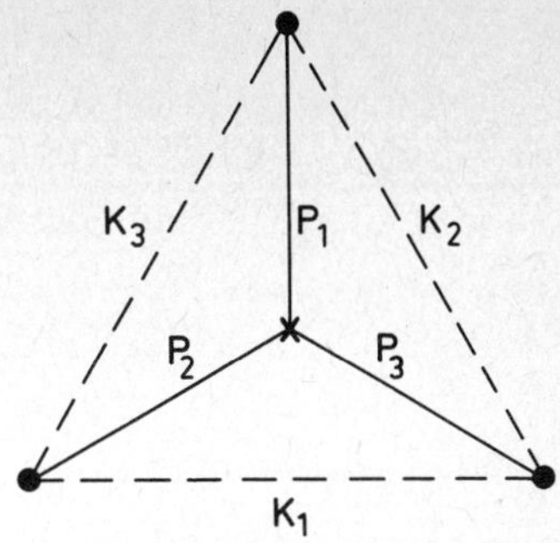

Fig.1.4. The star-triangle transformation. The three s spins (dots) are coupled to the central s' spin (cross) with coupling parameters p_1,p_2,p_3. Decimating the s' spin leads to interactions K_1,K_2,K_3 between the s spins given by (1.22,23)

Fig.1.3. The s and s' spin systems. The dots denote s spins and the crosses s' spins. The lines connecting the spins represent coupling parameters in $H_c(s',s)$

of s and s' spins are L and L-a, respectively, where a is the lattice constant of the system of s spins. In the limit $L \to \infty$, which will be taken below, the numbers of s and s' spins differ by a fraction of order a/L.

We now consider specifically the example of the Ising model. It will become clear that the same approach can also be applied to other two-dimensional models for which there is a star-triangle transformation that only produces nearest-neighbor interactions, for example, the Gaussian model. In the Ising case the coupling Hamiltonian $H_c(s',s)$ is chosen to be a sum of terms of the form $(p_1s_1 + p_2s_2 + p_3s_3)s'$, which couple each s' spin to the 3 neighboring s spins. In addition $H_c(s',s)$ contains an additive spin-independent term that must be considered in calculating the free energy but plays no role in the determination of the critical surface and critical exponents. The Hamiltonians $H'(s')$ and $H(s)$ are calculated from $H_c(s',s)$ in accordance with (1.20,21) by eliminating (decimating) the s and s' spins, respectively, with the star-triangle transformation [1.72-74]. Eliminating the s' spin at the center of a triangle of s spins, as in Fig.1.4, generates the nearest-neighbor couplings:

$$K_i = F(p_i,p_j,p_k) \quad , \tag{1.22}$$

$$F(p_i,p_j,p_k) = \frac{1}{4} \ln \frac{\cosh(p_i + p_j + p_k) \cosh(-p_i + p_j + p_k)}{\cosh(p_i - p_j + p_k) \cosh(p_i + p_j - p_k)} \quad , \tag{1.23}$$

where i,j,k take the values 1,2,3 and cyclic permutations thereof.

Note that each s' spin in Fig.1.3 is surrounded by 3 s spins, whereas the s spins at the edges of the equilateral triangle have only 2 neighboring s' spins. This asymmetry creates a spatial dependence of the renormalized coupling constants that works its way toward the interior of the triangle as the renormalization proceeds, even if the couplings in the initial Hamiltonian are spatially homogeneous. Thus we incorporate the possibility of spatially-dependent couplings in our analysis. In the limit $L \to \infty$ of an infinitesimal transformation the K's and p's are regarded as continuous functions of position.

The spatial dependence may be taken into account as follows. We denote by $K_i(\underline{R})$ and $p_i(\underline{R})$, $i = 1,2,3$, the coupling parameters (Figs.1.3,4) in the up-pointing triangle of s spins centered at $\underline{R}$. These quantities are related by (1.22,23). We denote the nearest-neighbor couplings in an up-pointing triangle of s' spins centered at $\underline{R}$ (Fig.1.5) by $\tilde{K}_i(\underline{R})$. Each of the $\tilde{K}_i$ is determined by a star-triangle transformation involving p_i in 3 different up-pointing triangles of s spins. The explicit relation has the form

$$\tilde{K}_i(\underline{R}) = F(p_i(\underline{R} + 2a\underline{e}_i),\ \ p_j(\underline{R} + a\underline{e}_i + a\underline{e}_j),\ \ p_k(\underline{R} + a\underline{e}_i + a\underline{e}_k)\ \ , \tag{1.24}$$

where $\underline{e}_1 = (0, -1/\sqrt{3})$, $\underline{e}_2 = (\frac{1}{2}, 1/2\sqrt{3})$, $\underline{e}_3 = (-\frac{1}{2}, 1/2\sqrt{3})$, as in Fig.1.5, and where i,j,k are cyclic. Assuming that the p_i only vary significantly over distances of order L, one may expand (1.24) in the form

$$\tilde{K}_i(\underline{R}) = K_i(\underline{R}) + a \sum_{\ell} Q_{i\ell}(\underline{e}_i + \underline{e}_\ell) \cdot \nabla p_\ell + O(a^2/L^2) \quad , \tag{1.25}$$

where $Q_{i\ell} = \partial F(p_i,p_j,p_k)/\partial p_\ell$. The ∇p_ℓ in (1.25) may be expressed in terms of the ∇K_j using the substitution $\nabla p_\ell = \sum_j Q^{-1}_{\ell j} \nabla K_j$, where Q^{-1} is the inverse of the matrix Q.

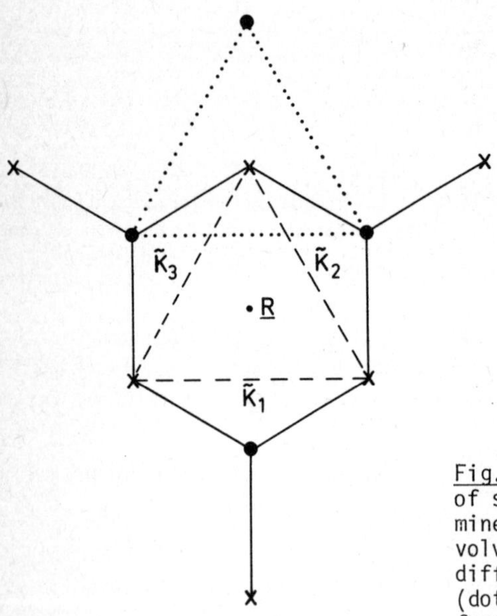

(a)

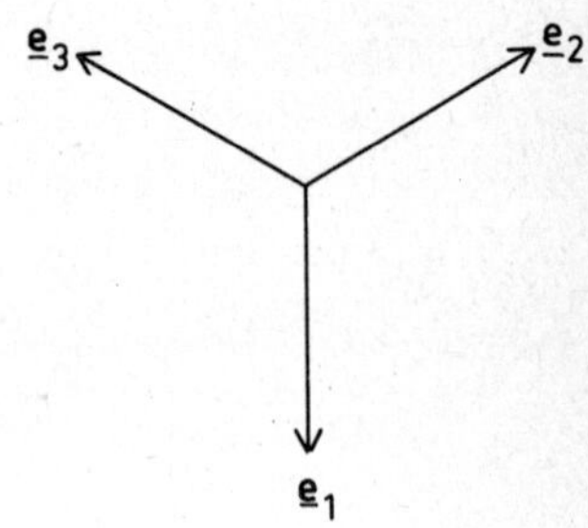

(b)

Fig.1.5. (a) Each of the $\tilde{K}_i$ for the triangle of s' spins (crosses) centered at $\underline{R}$ is determined by a star-triangle transformation involving the coupling parameters p_i in three different up-pointing triangles of s spins (dots). (b) The vectors $\underline{e}_1, \underline{e}_2, \underline{e}_3$ that point from an s spin (dot) to the three nearest s' spin (crosses)

The next step in deriving a renormalization transformation is to uniformly dilate the lattice of s' spins until the edges of the triangular lattices of s and s' spins in Fig.1.3 coincide. This may be achieved by choosing the origin of coordinates at the center of the triangular lattice of s' spins and multiplying distances to points on the lattice by L/(L-a). Defining renormalized couplings on the dilated lattice by $K_i'([L/(L-a)]\underline{R}) = \tilde{K}(\underline{R})$, expanding in a/L, substituting for $\tilde{K}(R)$ from (1.25), introducing $\underline{r} = \underline{R}/L$, setting $\delta K_i = K_i'(\underline{R}) - K_i(\underline{R})$ and $\delta t = a/L$, and taking the limit $L \to \infty$, one is led to the exact partial differential equation

$$\frac{\partial K_i}{\partial t} = \sum_j \underline{D}_{ij} \cdot \nabla K_j - \underline{r} \cdot \nabla K_i \quad , \tag{1.26}$$

where ∇ now indicates differentiation with respect to $\underline{r}$. The quantity $\underline{D}_{ij} = \sum_\ell (\underline{e}_i + \underline{e}_\ell) Q_{i\ell} Q_{\ell j}^{-1}$ is a highly nonlinear function of the K_i. Equation (1.26) is to be solved in an equilateral triangle with sides of length 1. In the star-triangle transformation that eliminates the s spins at the edges of the system in Fig.1.3, p_1, p_2, and p_3 vanish along the lower, right, and left sides, respectively. The solutions of (1.26) of interest satisfy this boundary condition or the equivalent condition $K_2 = K_3 = 0$, $K_3 = K_1 = 0$, $K_1 = K_2 = 0$ on the lower, right, and left edges, respectively, implied by (1.22,23).

We now summarize some of the important critical properties of (1.26). The reader interested in more details is referred to [1.17,18]. It can be shown that the transformation equation leaves the subspace of $K_i(\underline{r})$ that satisfy the criticality condition $\sum_i u_i(\underline{r}) = 1$, $u_i = \sinh 2K_j \sinh 2K_k$ (i,j,k cyclic) invariant. The criticality condition recalls Houtappel's [1.72] expression $\sum_i u_i = 1$ for the critical surface of the homogeneous Ising model on a triangular lattice with anisotropic couplings K_1, K_2, K_3. Within the invariant subspace there is a critical fixed point $K_i^*(\underline{r})$ that corresponds to $u_i^*(\underline{r}) = 1/3 - 2\underline{r} \cdot \underline{e}_i$. The couplings $K_i^*(\underline{r})$ are isotropic near the center but one-dimensional and infinitely strong at the edges. Finally we consider the flow equation obtained by linearizing (1.26) in $\psi_i(\underline{r}) = K_i(\underline{r}) - K_i^*(\underline{r})$ at the fixed point. One finds that any function $f(\underline{r})\xi_i(\underline{r})$, where $f(\underline{r})$ is arbitrary and $\xi_i(\underline{r})$ is the vector normal to the critical surface at the fixed point ($\xi_i(r) \propto \sum_j \partial u_j(\underline{r})/\partial K_i$), is an eigenfunction with eigenvalue $y_t = \nu^{-1} = 1$, in agreement with the known exact result for the thermal scaling index of the two-dimensional Ising model.

The free energy of the system for arbitrary coupling constants can, in principle, be calculated from the solution of the flow equation (1.26). However, because of the complicated nonlinear form of the partial differential equation, there is little hope of obtaining a general analytic solution. A special class of solutions is examined in [1.18].

So far no one has succeeded in calculating the magnetic scaling index y_h exactly by an extension of the method described above (see, however, [1.100]). If single-spin as well as two-spin couplings are included in $H_c(s',s)$, the star-triangle transformation produces one, two, and three-spin interactions in both H(s) and H'(s'). However, since the three spin-interactions only involve the s spins in up-pointing triangles and the s' spins in down-pointing triangles, the renormalization operation is no longer closed. A similar problem with three-spin interactions is encountered in the case of the Potts model.

1.6.2 Discussion

Differential renormalization transformations of the type considered here have also been devised for the one-dimensional Ising model [1.19] and for the two-, three-, and d-dimensional Gaussian models [1.20,21]. A prerequisite of the approach in its present form is a star-triangle transformation, i.e., a decimation that only produces nearest-neighbor interactions. The Ising model on a square lattice does not meet this requirement. Despite initial optimism concerning the possibility of exact differential renormalization for the three-dimensional Ising model (there are many three-dimensional lattices with coordination number 3), no successes have been reported thus far.

The exact differential renormalization transformations that have been constructed are of particular interest to persons concerned with fundamental mathematical aspects of the renormalization group. There has been little work to data on developing approximate differential real-space renormalization procedures (see, however, [1.99, 100]). The spatial dependence of the coupling constants is an inconvenient complication. However, only after dropping the natural prejudice against spatially dependent couplings has it thus far proved possible to renormalize the two-dimensional Ising model exactly.

DEKEYSER and STELLA [1.22] have discussed a differential renormalization of mean-field (van der Waals) models in which the spins interact with infinite-range constant infinitesimal couplings. The critical behavior of these models is obviously independent of the dimension and the lattice type, and the renormalization can be carried out without spatially-dependent couplings. The transformation has no Griffiths-Pearce singularities [1.14-16]. In general one would expect differential transformations to be free of such singularities, since only an infinitesimal fraction of the degrees of freedom are summed in the renormalization step. For the explicit $H_c(s',s)$ for the two-dimensional Ising model assumed above, it is clear that the sums in (1.20,21) are free of critical fluctuations.

1.7 Phenomenological Renormalization

1.7.1 Description of the Method

Phenomenological renormalization [1.23,75-78] is a type of finite-size scaling analysis invented by NIGHTINGALE [1.23]. We are including a discussion of the method in this article since it bears a certain resemblance to real-space renormalization, is applicable to the same kinds of systems, and is an extremely powerful calculational tool. It is difficult to make error estimates with conventional real-space renormalization methods. Systematically improving the approximations is possible in principle but difficult in practice. In phenomenological renormalization the approximation can actually be systematically improved many times in practice by increasing the finite size. Reliable extrapolations can be made by observing the convergence of the results, which is generally rapid.

To simplify the discussion of the method, we begin by considering a two-dimensional spin system, an Ising model, for example, which is characterized by two interaction constants: a nearest-neighbor ferromagnetic coupling K and a magnetic field H. A quantity of primary interest in phenomenological renormalization is the correlation length $\xi_\ell(K,H)$ in an infinitely long strip of spins ℓ spins wide. This correlation length, which characterizes the decay of the spin-spin correlation function parallel to the strip, can be expressed as

$$\xi_\ell^{-1} = \ln(\lambda_0^{(\ell)}/\lambda_1^{(\ell)}) \quad , \tag{1.27}$$

where $\lambda_0^{(\ell)}$ and $\lambda_1^{(\ell)}$ are the largest and second largest eigenvalues of the transfer matrix [1.79,80] that transfers interactions along the strip. A recurrence relation for the coupling constants is obtained in phenomenological renormalization by relating the correlation lengths in two strips with different widths ℓ and ℓ' according to

$$\xi_{\ell'}(K',H') = (\ell'/\ell)\xi_\ell(K,H) \quad . \tag{1.28}$$

This formula is justified below using both heuristic renormalization and finite-size scaling arguments. Predictions for the critical coupling and the thermal and magnetic critical exponents are obtained from (1.28) following the standard procedure of perturbing about a fixed point $K^* = K_c$, $H^* = 0$. The approximation may be improved by considerating larger ℓ and ℓ', as discussed below.

Equation (1.28) is suggested by KADANOFF's intuitive picture [1.24], according to which cells of b^d spins in a system near the critical point behave like the individual spins in a similar system farther from criticality. The absolute correlation length is the same in the two equivalent descriptions. Thus the correlation lengths per lattice constant in the site and cell systems are related by

$$\xi(K',H') = b^{-1}\xi(K,H) \quad . \tag{1.29}$$

Equation (1.28) postulates a similar relationship between the correlation lengths in an infinite strip with coupling constants K, H and width ℓ and an equivalent infinite strip with renormalized couplings K',H' and width $\ell' = \ell/b$. Neither KADANOFF's intuitive picture nor the fundamental equation (1.28) of phenomenological renormalization consider the possibility of proliferation of interactions, i.e., in writing (1.28,29) it is not deemed necessary to expand the space of coupling constants to include other couplings besides K and H.

Phenomenological renormalization may also be interpreted as a type of finite-size scaling analysis without recourse to renormalization-group ideas [1.75-78]. According to a basic postulate of finite-size scaling [1.81], in a strip of width $\ell \gg 1$ with couplings such that $\xi_\ell >> 1$, ξ_ℓ and ξ_∞ are related by

$$\xi_\ell(K,H) = \xi_\infty(K,H)Y(\ell/\xi_\infty(K,H)) \quad . \tag{1.30}$$

The scaling function Y(x) approaches the value 1 for $x \gg 1$, so that (1.30) reduces to an identity in the limit $\ell \to \infty$. Y(x) behaves as Ax for $0 < x \ll 1$, so that ξ_ℓ remains finite as ξ_∞ diverges in the limit $K \to K_c$, $H \to 0$. From (1.30) and the asymptotic form of Y(x) for $x \ll 1$, one finds $\xi_\ell(K_c,0) = A\ell$, which implies

$$\xi_{\ell'}(K_c,0) = (\ell'/\ell)\xi_\ell(K_c,0) \tag{1.31}$$

for $\ell,\ell' \gg 1$. Thus finite-size scaling also implies that $K = K_c$, $H = 0$ is a fixed point of (1.28).

To derive the recurrence relation (1.28) from the finite-size scaling hypothesis, we use (1.30) to write

$$\xi_{\ell'}(K',H') = \frac{\xi_\infty(K',H')Y(\ell'/\xi_\infty(K',H'))}{\xi_\infty(K,H)Y(\ell/\xi_\infty(K,H))} \xi_\ell(K,H) \quad . \tag{1.32}$$

If for a given K,H,ℓ, and ℓ', K' and H' are chosen so that

$$\xi_\infty(K',H') = (\ell'/\ell)\xi_\infty(K,H) \quad , \tag{1.33}$$

the unknown function Y drops out of (1.32), and the recurrence relation (1.28) of phenomenological renormalization is obtained.

According to finite-size scaling, (1.28,33) are satisfied simultaneously. The procedure to be followed in calculating critical indices with (1.28) follows from (1.33). The correlation length in the infinite system [1.1-9,24] diverges as $|K-K_c|^{-1/y_t}$ for H = 0 and as $|H|^{-1/y_h}$ for $K = K_c$. Thus the K' and H' defined by (1.33) satisfy the same transformation equations [1.1-9,24]

$$K' - K_c = (\ell/\ell')^{y_t}(K - K_c) \quad , \quad H = 0 \tag{1.34}$$

$$H' = (\ell/\ell')^{y_h} H \quad , \quad K = K_c \tag{1.35}$$

as the relevant thermal and magnetic scaling fields in a renormalization transformation with length-rescaling factor $b = \ell/\ell'$. Since (1.32,33) imply (1.28), predictions for the critical indices may also be obtained by analyzing the fundamental recurrence relation (1.28) of phenomenological renormalization, (1.34), and (1.35) simultaneously, following the standard procedure in renormalization-group calculations.

Phenomenological renormalization, unlike ordinary renormalization-group procedures, does not furnish as many recurrence relations as there are coupling constants. The single recurrence relation (1.28) does not determine K' and H' uniquely for a given K and H. In this particular example it is clear how to calculate two critical exponents with a single recurrence relation. Since the paramagnetic-ferromagnetic phase transition only takes place with $H=0$, i.e., since $K-K_c$ and H are independent relevant variables, one looks for a fixed point with $K^*=K_c$, $H^*=0$ and linearizes (1.28) independently in K and H, as in (1.34,35). (Actually one linearizes $\xi_\ell(K,H)$ in H^2, not H, since $\xi_\ell(K,H)=\xi_\ell(K,-H)$ is analytic in H for finite ℓ).

It is less obvious how to proceed if the Hamiltonian contains couplings that change the critical temperature but not the universality class. We now briefly consider such an example, a spin system with ferromagnetic first- and second-neighbor couplings in the absence of a magnetic field. Phenomenological renormalization yields the recurrence relation

$$\xi_{\ell'}(K_1',K_2') = (\ell'/\ell)\xi_\ell(K_1,K_2) \quad , \tag{1.36}$$

which in general has a fixed line $K_1^*(K_2)$. A finite-size scaling analyis based on the analog of (1.30) indicates that as $\ell,\ell' \to \infty$, the fixed line converges to the exact critical line. Note that the entire critical line is not mapped onto a single fixed point, as in ordinary renormalization-group methods.

When (1.36) is linearized about a point $\underline{K}^*$ on the fixed line, one obtains

$$[\nabla_{\underline{K}'}\xi_{\ell'}(\underline{K}^*)]\cdot\underline{\delta K}' = (\ell'/\ell)[\nabla_{\underline{K}}\xi_\ell(\underline{K}^*)]\cdot\underline{\delta K} \quad , \tag{1.37}$$

where $\underline{K}=(K_1,K_2)$ and $\underline{\delta K}=\underline{K}-\underline{K}^*$. In the limit $\ell,\ell' \gg 1$, the two gradients in (1.37) are parallel, and any vector $\underline{\delta K}$ is an eigenvector such that $\underline{\delta K}'=(\ell/\ell')^{y_t}\,\underline{\delta K}$, with

$$(\ell/\ell')^{y_t+1} = \frac{|\nabla_{\underline{K}}\xi_\ell(\underline{K}^*)|}{|\nabla_{\underline{K}'}\xi_{\ell'}(\underline{K}^*)|} \quad . \tag{1.38}$$

For finite ℓ and ℓ', on the other hand, the two gradients are not parallel, in general, and (1.37) does not determine the eigenvalue uniquely. To proceed, one usually specifies in addition that the eigenvector be parallel to a fixed, but arbitrary, vector $\underline{n}$, so that (1.38) is replaced by

$$(\ell/\ell')^{y_t+1} = \frac{\underline{n}\cdot\nabla_{\underline{K}}\xi_\ell(\underline{K}^*)}{\underline{n}\cdot\nabla_{\underline{K}'}\xi_{\ell'}(\underline{K}^*)} \quad . \tag{1.39}$$

With a finite-size scaling analysis similar to that given above, one can show that in the limit $\ell,\ell' \to \infty$ the y_t defined by (1.39) converges to the exact value for any $\underline{n}$ and $\underline{K}^*$ as long as $\underline{n}$ is not parallel to the critical line at point $\underline{K}^*$. The rate of convergence depends, of course, on $\underline{K}^*$ and $\underline{n}$. KINZEL and SCHICK [1.82] have shown in a particular example that the convergence is particularly rapid at the point on the fixed line where the two gradients are parallel, i.e., where the prediction for y_t is independent of $\underline{n}$.

Phenomenological renormalization becomes asymptotically exact in the limit $\ell,\ell' \gg 1$, in which the finite-size scaling hypothesis (1.30) holds. In applications of the method it is important to choose ℓ and ℓ' as large as possible. The largest ℓ' for a given ℓ is, of course, $\ell' = \ell - 1$. Considerable effort must be invested in devising numerical procedures for calculating the eigenvalues $\lambda_0^{(\ell)}$ and $\lambda_1^{(\ell)}$ in (1.27) for large values of ℓ because of the enormous dimension of the transfer matrix. Useful simplifications can be achieved by using periodic or helical boundary conditions, group-theoretical reduction procedures, and sparse-matrix techniques [1.23,82-86]. In applications to two-dimensional Ising-like models, maximum strip widths of about 16 have been attained [1.23].

In devising effective extrapolation procedures, it would be useful to know how the deviations of the predictions of phenomenological renormalization from the exact results tend toward zero as ℓ and ℓ' increase. No general results are available so far (see, however [1.86,87]), but in the case of an Ising strip with periodic boundary conditions DERRIDA and VANNIMENUS [1.75,76] have proved that

$$K_c(\ell,\ell') = \frac{1}{2}\ln(1+\sqrt{2}) - \frac{\pi^3}{192}\frac{b(b+1)}{\ell^3} + \dots \quad , \tag{1.40}$$

$$\nu(\ell,\ell') = 1 - \frac{\pi^2 \ln 2}{24}\frac{b^2-1}{\ell^2 \ln b} + \dots \quad , \tag{1.41}$$

where K_c is the critical nearest-neighbor coupling, the critical exponent $\nu = y_t^{-1}$, and $b = \ell/\ell'$. One sees that the convergence to the exact results is rapid and that the optimal choice for ℓ' is $\ell - 1$.

In order to examine the connection between ordinary microscopic renormalization, which generally is beset by the proliferation problem, and phenomenological renormalization, which is not, we consider a microscopic renormalization transformation that maps a system with nearest-neighbor coupling K_1 and linear dimension ℓ onto a system with couplings $\tilde{K}_1,\tilde{K}_2,\dots$ and linear dimension ℓ'. The correlation lengths of the two systems are related by

$$\xi_{\ell'}(\tilde{K}_1,\tilde{K}_2,\dots) = (\ell'/\ell)\xi_\ell(K_1,0,\dots) \quad , \tag{1.42}$$

which is a generalization of (1.28) to a many-parameter space. We then search for a point $K_1',0,\dots$ such that

$$\xi_{\ell'}(K_1',0,\ldots) = \xi_{\ell'}(\tilde{K}_1,\tilde{K}_2,\ldots) \quad , \tag{1.43}$$

i.e., we supplement the renormalization by a second mapping with no change in length scale to a system with nearest-neighbor interactions only which has the same correlation length. Combining these two operations yields a phenomenological renormalization transformation

$$\xi_{\ell'}(K_1',0,\ldots) = (\ell'/\ell)\xi_{\ell}(K_1,0,\ldots) \tag{1.44}$$

which only involves a single coupling constant.

Note that (1.44) fails to recognize the irrelevant variables that determine corrections to scaling, unless the nearest-neighbor coupling happens to be an exact scaling field of the particular microscopic renormalization transformation. This is an inherent limitation of the phenomenological renormalization approach, which, however, is not too serious if one is only interested in calculating the leading singular behavior.

That the single recurrence relation of phenomenological renormalization, for example, (1.36), has a fixed line or surface in the case of systems with several coupling constants makes the method particularly suitable for applications to systems with nonuniversal behavior [1.23,78], where the critical exponents vary on the critical surface. The approximations in standard real-space renormalization techniques generally eliminate all trace of nonuniversal behavior. In mapping out the global phase diagram of a system with several different types of phase transitions, such as the Blume-Emery-Griffiths model [1.88], the standard renormalization techniques, which project entire critical surfaces onto single fixed points, have the advantage of predicting precise phase boundaries without complicated calculations. In phenomenological renormalization, the predicted critical exponents vary weakly on critical surfaces where the exact exponent is everywhere the same. Phase boundaries are located where the exponents exhibit discontinuities smoothed by finite-size effects.

The finite-size scaling hypothesis (1.30) only holds if the two important physical lengths, ℓ and ξ, are very large in comparison with the lattice constant. Phenomenological renormalization is particularly suited to the study of phase transitions in which the correlation length diverges. It is not clear that one can expect useful information in the case of first-order transitions in which the condition $\xi \gg 1$ is never fulfilled. In conventional microscopic real-space renormalization first-order transitions are associated, in principle, with special discontinuity fixed points [1.12,13].

1.7.2 Applications

In [1.23,82,89-92] applications of phenomenological renormalization to critical phenomena in classical systems are described. In [1.78,93,94] the method is applied to quantum Hamiltonian systems. Percolation is considered in [1.75,76,86,95,96], the self-avoiding walk in [1.97], and Anderson localization in [1.98]. Most of these papers consider two-dimensional systems (see, however [1.98]), but the method is also applicable, in principle, in higher dimensions.

1.8 Concluding Remarks

This completes our review of developments in real-space renormalization. The approach provides a versatile and intuitive means to study phase transitions in low-dimensional systems with simple degrees of freedom. The most difficult part of the approach is actually devising the procedure for eliminating a fraction of the degrees of freedom. Choosing a suitable weight function requires some insight into the physics driving the phase transition, since the weight function should suppress the critical fluctuations that lead to singularities in the free energy. In general, the renormalization cannot be carried out exactly, and a systematic expansion parameter has not been found. (The inverse range of the potential [1.55] and the number of states of the Potts model [1.99] have been suggested as possible candidates).

Several new methods (Sect.1.3) for performing an approximate elimination of degrees of freedom have appeared in the last five or six years, each of which has its advantages and limitations. In the case of the two-dimensional Ising model on a triangular lattice and a few other models with very special properties (Sect.1.6), exact differential renormalization procedures have been devised. A powerful phenomenological renormalization approach (Sect.1.7) has been developed, which makes no attempt to eliminate degrees of freedom on a microscopic level but extracts information from a scaling relation between the correlation lengths in systems with different finite sizes.

Only limited progress has been made in dealing with a number of fundamental inherent problems (Sect.1.5) in real-space renormalization. One such problem is the infinite proliferation of interactions, which arises with most of the weight functions used so far but is truncated by the approximation schemes. In the case of exact differential transformations this problem is avoided. Phenomenological renormalization goes to the extreme of allowing only the temperature and the magnetic field as renormalization parameters. However, in the case of the Potts model [1.37-39], where it is important to include vacancies in describing the change in the order of the transition (Sect.1.5.1), this strategy is not completely successful [1.91].

A second, closely-related problem is the ambiguity in the choice of the weight function (Sect.1.5.1). Whether fixed points with all the properties one normally assumes really exist in exact transformations with the usual weight functions, i.e., whether the freedom in the choice of the weight function granted by (1.4) is real or deceptive, is a question about which little is known. From the work of GRIFFITHS and PEARCE (Sect.1.5.2) it has become fairly clear that exact transformations with most of these weight functions would not be globally nonsingular in the coupling constants. However, no such singularities have so far been found in the critical region, and differential renormalization procedures seem to be less prone to this particular difficulty. The importance of including vacancies in renormalizing the Potts model (Sect.1.5.1) indicates clearly that selecting the renormalized spin variables and the weight function is a subtle matter requiring physical insight and that renormalization involves much more than just "turning the crank". In the present state of the art choosing the renormalized variables and the weight function for an approximate calculation is somewhat like choosing the trial wavefunction for a variational calculation in quantum mechanics. The predictions depend on the choice, and making a satisfactory choice requires a certain amount of physical insight.

A third fundamental problem is the difficulty of estimating the error and making systematic improvements in the relatively crude approximations on which many of the renormalization procedures are based. Monte Carlo and phenomenological renormalization succeed to some extent in dealing with this problem.

During the preparation of this article the list of references was amended several times to include work that had just appeared. Significant developments in real-space renormalization and its applications are continuing to take place and can be anticipated in the future. We hope that this and the other chapters in this volume provide useful summaries of what has been accomplished thus far.

Acknowledgements. We have benefitted from stimulating and informative discussions with a large number of colleagues over a period of many years. In particular we wish to thank H.W.J. Blöte, B. Derrida, P.C. Hemmer, H.J. Hilhorst, H.J.F. Knops, S.K. Ma, M.P. Nightingale, M.P.M. den Nijs, A.M.M. Pruisken, W. van Saarloos, M. Schick, and D.J. Wallace.

References

1.1 C. Domb, M.S. Green (eds.): *Phase Transitions and Critical Phenomena*, Vol.6 (Academic, New York 1976)
1.2 K.G. Wilson, J. Kogut: Phys. Rep. *12*C, 75 (1974)
1.3 M.E. Fisher: Rev. Mod. Phys. *46*, 597 (1974)
1.4 K.G. Wilson: Rev. Mod. Phys. *47*, 773 (1975)
1.5 M.N. Barber: Phys. Rep. *29*C, 1 (1977)
1.6 D.J. Wallace, R.K.P. Zia: Rep. Prog. Phys. *41*, 1 (1978)
1.7 S. Ma: *Modern Theory of Critical Phenomena* (Benjamin, New York 1976)
1.8 P. Pfeuty, G. Toulouse: *Introduction to the Renormalization Group and to Critical Phenomena* (Wiley, New York 1977)
1.9 D.J. Amit: *Field Theory, the Renormalization Group, and Critical Phenomena* (McGraw-Hill, New York 1978)
1.10 J.C. Le Guillou, J. Zinn-Justin: Phys. Rev. Lett. *39*, 95 (1977); Phys. Rev. B*21*, 3976 (1980)

1.11 P.C. Hohenberg, B.I. Halperin: Rev. Mod. Phys. *49*, 435 (1977)
1.12 B. Nienhuis, M. Nauenberg: Phys. Rev. Lett. *35*, 477 (1975)
1.13 Th. Niemeijer, J.M.J. van Leeuwen: In [Ref.1.1, pp.425-505]
1.14 R.B. Griffiths, P.A. Pearce: Phys. Rev. Lett. *41*, 917 (1978); J. Stat. Phys. , 499 (1979)
1.15 R.B. Griffiths: In *Proc. Colloq. on Random Fields, Esztergom, Hungary, 1979*, to be published; Physica *106*A, 59 (1981)
1.16 R.B. Israel: In *Proc. Colloq. on Random Fields, Esztergom, Hungary, 1979*, to be published
1.17 H.J. Hilhorst, M. Schick, J.M.J. van Leeuwen: Phys. Rev. Lett. *40*, 1605 (1978); Phys. Rev. B*19*, 2749 (1979)
1.18 H.J.F. Knops, H.J. Hilhorst: Phys. Rev. B*19*, 3689 (1979)
1.19 W. van Saarloos, J.M.J. van Leeuwen, A.L. Stella: Physica *97*A, 319 (1979)
1.20 Y. Yamazaki, H.J. Hilhorst: Phys. Lett. *70*A, 329 (1979)
1.21 Y. Yamazaki, G. Meissner, H.J. Hilhorst: Z. Phys. B*35*, 333 (1979); Y. Yamazaki, H.J. Hilhorst, G. Meissner: J. Stat. Phys. *23*, 609 (1980)
1.22 R. Dekeyser, A. Stella: J. Stat. Phys. *23*, 587 (1980)
1.23 M.P. Nightingale: Physica *83*A, 561 (1976); Phys. Lett. *59*A, 486 (1977); Proc. K. Ned. Akad. We. Ser. B *82* (3), 235 (1979)
1.24 L.P. Kadanoff, W. Götze, D. Hamblen, R. Hecht, E.A.S. Lewis, V.V. Palciauskas, M. Rayl, J. Swift, D. Aspnes, J. Kane: Rev. Mod. Phys. *39*, 395 (1967)
1.25 D.M. Burley: In *Phase Transitions and Critical Phenomena*, ed. by C. Domb, M.S. Green, Vol.2 (Academic, New York 1972)
1.26 L.P. Kadanoff: Phys. Rev. Lett. *34*, 1005 (1975)
1.27 L.P. Kadanoff, A. Houghton, M.C. Yalabik: J. Stat. Phys. *14*, 171 (1976)
1.28 W. van Saarloos, J.M.J. van Leeuwen, A.M.M. Pruisken: Physica *92*A, 323 (1978)
1.29 M.P.M. den Nijs, H.J.F. Knops: Physica *93*A, 441 (1978)
1.30 A.A. Migdal: Sov. Phys. JETP *42*, 413, 743 (1976)
1.31 L.P. Kadanoff: Ann. Phys. (N.Y.) *100*, 359 (1976)
1.32 S. Ma: Phys. Rev. Lett. *37*, 461 (1976)
1.33 Z. Friedman, J. Felsteiner: Phys. Rev. B*15*, 5317 (1977)
1.34 A.L. Lewis: Phys. Rev. B*16*, 1249 (1977)
1.35 R.H. Swendsen: Phys. Rev. Lett. *42*, 859 (1979)
1.36 K. Binder: Phys. Rev. Lett. *47*, 693 (1981)
1.37 B. Nienhuis, A.N. Berker, E.K. Riedel, M. Schick: Phys. Rev. Lett. *43*, 737 (1979)
1.38 B. Nienhuis, E.K. Riedel, M. Schick: J. Phys. A*13*, L31, L189 (1980); Phys. Rev. B*23*, 6055 (1981)
1.39 T.W. Burkhardt: Z. Phys. B*39*, 159 (1980)
1.40 M. Suzuki: Prog. Theor. Phys. *46*, 1337 (1971)
1.41 M. Suzuki: Prog. Theor. Phys. *56*, 1454 (1976)
1.42 P.W. Kasteleyn, C.M. Fortuin: J. Phys. Soc. Jpn. Suppl. *26*, 11 (1969)
1.43 P.G. de Gennes: Phys. Lett. *38*A, 339 (1972)
1.44 U. Deker, F. Haake: Z. Phys. B*36*, 379 (1980)
1.45 J.M. Kosterlitz, D.J. Thouless: J. Phys. C*6*, 1181 (1973)
1.46 J.M. Kosterlitz: J. Phys. C*7*, 1046 (1974)
1.47 J.V. José, L.P. Kadanoff, S. Kirkpatrick, D.R. Nelson: Phys. Rev. B*16*, 1217 (1977)
1.48 D.J. Amit, Y.Y. Goldschmidt, G. Grinstein: J. Phys. A*13*, 585 (1980)
1.49 H.J.F. Knops: Ann. Phys. (N.Y.) *128*, 448 (1980)
1.50 J.L. Black, V.J. Emery: Phys. Rev. B*23*, 429 (1981)
1.51 K.G. Wilson: Phys. Rep. *23*C, 331 (1976)
1.52 L.P. Kadanoff: In *Critical Phenomena*, ed. by J. Brey, R.B. Jones (Springer, Berlin, Heidelberg, New York 1976) pp.276-346
1.53 S.D. Drell, B. Svetitsky, M. Weinstein: Phys. Rev. D*17*, 523 (1978)
1.54 T.W. Burkhardt, W. Kinzel: Phys. Rev. B*20*, 4730 (1979)
1.55 H.J.F. Knops, J.M.J. van Leeuwen, P.C. Hemmer: J. Stat. Phys. *17*, 197 (1977)
1.56 J.J.M. Rijpkema, H.J.F. Knops: Physica *108* A, 589 (1981)
1.57 T.W. Burkhardt, E. Eisenriegler: Phys. Rev. B*16*, 3213 (1977)
1.58 G.F. Mazenko, J.E. Hirsch, M.J. Nolan, O.T. Valls: Phys. Rev. B*23*, 1431 (1981)
1.59 L.P. Kadanoff, A. Houghton: Phys. Rev. B*11*, 377 (1975)

1.60 J.M.J. van Leeuwen: Phys. Rev. Lett. *34*, 1056 (1975)
1.61 A.L. Stella, R. Dekeyser, J. Indekeu: Phys. Rev. B*20*, 3799 (1979)
1.62 A.N. Berker: Phys. Rev. B*12*, 2752 (1975)
1.63 T.W. Burkhardt, R.H. Swendsen: Phys. Rev. B*13*, 3071 (1976)
1.64 J. Sòlyom, P. Pfeuty: Phys. Rev. B*24*, 218 (1981)
1.65 D. Andelman, A.N. Berker: J. Phys. A*14*, L91 (1981)
1.66 A.N. Berker, D. Andelman, A. Aharony: J. Phys. A*13*, L413 (1980)
1.67 H.J.F. Knops: Physica *93*A, 427 (1978)
1.68 T.W. Burkhardt: Phys. Rev. Lett. *43*, 1629 (1979)
1.69 Y. Imry, S. Ma: Phys. Rev. Lett. *35*, 1399 (1975)
1.70 E. Pytte, Y. Imry, D. Mukamel: Phys. Rev. Lett. *46*, 1173 (1981); K. Binder, Y. Imry, E. Pytte: Phys. Rev. B*24*, 6736 (1981)
1.71 A. Stella: Physica *108* A, 211 (1981)
1.72 R.M.F. Houtappel: Physica (Utrecht) *16*, 425 (1950)
1.73 G.F. Newell: Phys. Rev. *79*, 876 (1950)
1.74 I. Syozi: In *Phase Transitions and Critical Phenomena,* Vol.1, ed. by C. Domb, M.S. Green (Academic, New York 1972)
1.75 B. Derrida, J. Vannimenus: In *Colloque sur les méthodes de calcul pour l'étude de phénomènes critiques de Carry le Rouet, 1980* (Springer, Berlin, Heidelberg, New York)
1.76 B. Derrida, L. de Seze, J. Vannimenus: In *Proc. Int. Conf. on Disordered Systems and Localization, Rome, 1981,* to be published
1.77 R.R. dos Santos, L. Sneddon: Phys. Rev. B*23*, 3541 (1981)
1.78 C.J. Hamer, M.N. Barber: J. Phys. A*14*, 241, 259 (1981)
1.79 H.A. Kramers, G.H. Wannier: Phys. Rev. *60*, 252 (1941)
1.80 C. Domb: Adv. Phys. *9*, 149 (1960)
1.81 M.E. Fisher, M.N. Barber: Phys. Rev. Lett. *28*, 1516 (1972)
1.82 W. Kinzel, M. Schick: Phys. Rev. B*23*, 3435 (1981); Phys. Rev. B*24*, 2701 (1981)
1.83 F.H. Ree, D.A. Chestnut: Phys. Rev. *45*, 3983 (1966)
1.84 L.K. Runnels, L.L. Combs: J. Chem. Phys. *45*, 2482 (1966)
1.85 C. Domb: Proc. Royal Soc. London A*196*, 36 (1949)
1.86 W. Kinzel, J. Yeomans: J. Phys. A*14*, L163 (1981)
1.87 M.N. Barber, M.E. Fisher: Ann. Phys. (N.Y.) *77*, 1 (1973)
1.88 A.N. Berker, M. Wortis: Phys. Rev. B*14*, 4946 (1976)
1.89 L. Sneddon: J. Phys. C*11*, 2823 (1978); J. Phys. C*12*, 3051 (1979)
1.90 Z. Rácz: Phys. Rev. B*21*, 4012 (1980)
1.91 H.W.J. Blöte, M.P. Nightingale, B. Derrida: J. Phys. A*14*, L45 (1981)
1.92 K. Binder: Z. Phys. B*43*, 119 (1981)
1.93 K. Uzelac, R. Jullien: J. Phys. A*14*, L151 (1981)
1.94 R.R. dos Santos, R.B. Stinchcombe: J. Phys. A*14*, 2741 (1981)
1.95 B. Derrida, J. Vannimenus: J. Phys. Lett. *41*, 473 (1980)
1.96 E. Domany, W. Kinzel: Phys. Rev. Lett. *47*, 5 (1981)
1.97 B. Derrida: J. Phys. A*14*, L5 (1981)
1.98 J.L. Pichard, G. Sarma: J. Phys. C*14*, L127 (1981)
1.99 H.J. Hilhorst, J.M.J. van Leeuwen: Physica *106*A, 301 (1981)
1.100 B. Payandeh, J.M.J. van Leeuwen: Physica *106*A, 589 (1981)

2. Bond-Moving and Variational Methods in Real-Space Renormalization

T. W. Burkhardt

With 6 Figures

A review of bond-moving and variational approximation methods in real-space renormalization is presented. Some of the more important developments in this field are illustrated in detail, using the Ising model as an example. Particular emphasis is placed on the Migdal-Kadanoff transformation and the Kadanoff lower-bound variational transformation. The successes and inherent limitations of the bond-moving and variational approximations are discussed. References to various applications of the methods are given.

2.1 Introduction

The partition function of a system of spin variables σ with Boltzmann factor $\exp[H(\sigma)]$ is given by

$$Z[H] = \sum_{\sigma} e^{H(\sigma)} \quad , \tag{2.1}$$

where the sum is over all configurations of the spins. In the real-space renormalization-group approach, which is reviewed in [2.1] and in Chap.1, the partition function is evaluated in steps. In each step the system of spins is replaced by an equivalent system with fewer spins. Formally the renormalization step is represented by the equation

$$e^{H'(\sigma')} = \sum_{\sigma} P(\sigma',\sigma)e^{H(\sigma)} \quad , \tag{2.2}$$

where H and H', two successive Hamiltonians in the sequence, describe systems of N and N' spins with $N' < N$. The weight function or projection operator $P(\sigma',\sigma)$, which relates the σ' and σ spins, satisfies the condition

$$\sum_{\sigma'} P(\sigma',\sigma) = 1 \quad , \tag{2.3}$$

so that the partition functions corresponding to H and H' are the same.

In general, the sum in (2.2) cannot be evaluated exactly. The first attempts [2.1] to apply (2.2) to spin systems on a lattice were made by NIEMEIJER and VAN LEEUWEN around 1973-1974 using cumulant and cluster approximations. Some of the other approximate renormalization methods that have been developed are reviewed in Chap.1 of this volume. Here in Chap.2 a class of real-space renormalization methods built around bond-moving and variational approximations is considered. The work to be described is largely based on ideas of KADANOFF and of MIGDAL.

In fundamental papers by KADANOFF and co-workers [2.2,3] which appeared in 1975 and 1976, variational and bond-moving approximation techniques were applied to real-space renormalization for the first time. Variational principles for the free energy provide criteria for optimizing certain approximate renormalization procedures. KADANOFF pointed out that the free energy in the first-order cumulant approximation is an upper bound to the exact free energy. He also showed that a class of renormalization transformations involving bond-moving approximations yields lower bounds to the free energy. With a simple and ingenious lower-bound transformation containing a variational parameter in the weight function, he obtained extremely accurate values for the critical exponents of the d = 2 Ising model and good results in higher dimensions as well.

The success of the lower-bound transformation is marred by an inconsistency that KADANOFF himself pointed out and which was studied in more detail by VAN SAARLOOS et al. [2.4] and DEN NIJS and KNOPS [2.5] (see also BARBER [2.6]). The best bound to the free energy is obtained if the variational parameter is allowed to change in each application of the renormalization transformation, whereas the highly accurate predictions of critical exponents were calculated with the parameter fixed at the optimal value at the fixed point. The analyses in [2.4,5] reveal some rather disturbing features common to approximate renormalization procedures based on variational principles for the free energy. In general, for the sequence of variational parameters that optimizes the free energy, the renormalization transformation is nonanalytic in the coupling constants at the fixed point and incapable of predicting a positive specific heat exponent α. Since KADANOFF's original work on the Ising model, critical exponents for a variety of other systems [2.3,7-32] have also been calculated with astonishing accuracy using the Kadanoff variational method with the variational parameter chosen according to KADANOFF's original prescription. Why this procedure works so well is still not well understood.

In 1975 MIGDAL [2.33] proposed an approximate renormalization scheme that has become one of the most widely used real-space renormalization procedures. It only involves nearest-neighbor couplings, can readily be continued to arbitrary dimension, and is applicable to systems with continuous as well as discrete spins and to lattice gauge theories [2.34,35]. KADANOFF [2.35] pointed out that MIGDAL's transformation, which will be referred to below as the Migdal-Kadanoff transformation, is a lower-bound transformation, since the renormalization step is equivalent to bond moving followed by one-dimensional decimation. The Migdal-Kadanoff transformation preserves the duality property of several two-dimensional models and thus yields exact critical surfaces, even though the corresponding critical exponents are sometimes unimpressive. The bond-moving approximation is exact at zero temperature. A number of predictions of the Migdal-Kadanoff transformation are exact to order ε in dimension $d = d_{lc} + \varepsilon$, where d_{lc} is the lower critical dimensionality at which the critical temperature of the system goes to zero.

In this chapter the bond-moving and variational approaches in real-space renormalization are reviewed. The Migdal-Kadanoff and Kadanoff lower-bound variational transformations are described at an introductory level using the Ising model as an example. Some other bond-moving and variational transformations are discussed briefly. The fundamental difficulties with the variational approach analyzed in [2.4,5] and alluded to above are outlined. Applications to a number of systems besides the Ising model are referenced but not considered in detail.

The article is organized as follows. Section 2.2 contains derivations of the variational principles for the free energy on which existing variational approximations in real-space renormalization are based. The Migdal-Kadanoff transformation and various modifications are discussed in Sect.2.3. Section 2.4 is devoted to the Kadanoff lower-bound variational transformation and other variational methods. Section 2.5 consists of a summary and concluding remarks.

2.2 Variational Principles

2.2.1 Lower-Bound Property of Bond-Moving Approximations

The sum over σ configurations in the exact renormalization prescription (2.2) is generally intractable. In lower-bound transformations of the type invented by KADANOFF [2.2,3], one replaces $H(\sigma)$ in (2.2) by $H(\sigma) + V(\sigma)$, with $V(\sigma)$ chosen so that the sum over σ configurations can be evaluated explicitly. Equation (2.2) is replaced by the approximate renormalization transformation

$$e^{H'(\sigma')} = \sum_{\sigma} P(\sigma',\sigma) e^{H(\sigma)+V(\sigma)} \quad . \tag{2.4}$$

For simplicity only classical spin systems, for which $H(\sigma)$ and $V(\sigma)$ commute, will be considered here.

Exact renormalization transformations of the form (2.2) leave the partition function invariant, whereas approximate transformations of the form (2.4) do not. Summing (2.4) over all σ' configurations and using property (2.3) and the inequality $\exp x \geqslant 1 + x$, one finds

$$\sum_{\sigma'} e^{H'(\sigma')} = \sum_{\sigma} e^{H(\sigma)+V(\sigma)} \geqslant \sum_{\sigma} e^{H(\sigma)}[1 + V(\sigma)] \tag{2.5}$$

or $Z[H'] = Z[H + V] \geqslant Z[H](1 + \langle V\rangle_H)$. Suppose that $\langle V\rangle_H = 0$. Then $Z[H'] \geqslant Z[H]$, or $F[H'] \leqslant F[H]$, where the free energy F is defined by $F = -\ln(Z)$, i.e., the free energy of the system with Hamiltonian H' is a lower bound[1] to the free energy of

(Footnote 1 see next page)

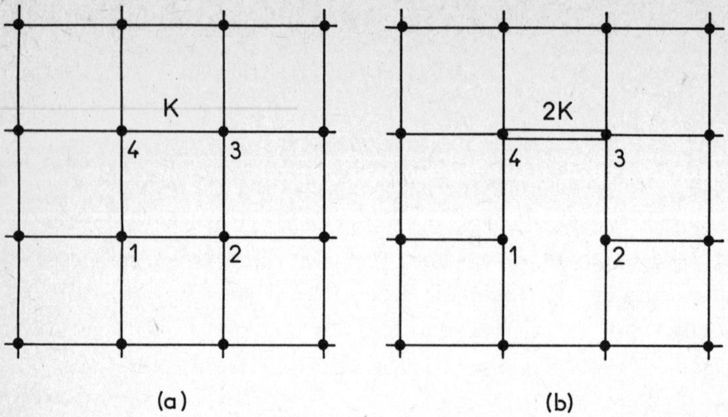

Fig.2.1a,b. Systems with Hamiltonians H and H+V, respectively. Adding V to H shifts the bond between spins 1 and 2 to spins 3 and 4

the system with Hamiltonian H. If an infinite sequence of transformations of the form (2.4) is carried out, and if in each step, for example, $H^{(n)} \rightarrow H^{(n+1)}$, $V^{(n)}$ is chosen so that $\langle V^{(n)} \rangle_{H^{(n)}} = 0$, a lower bound to the exact free energy corresponding to the initial Hamiltonian is obtained. In lower-bound variational calculations, variational parameters in the weight function $P(\sigma',\sigma)$ are adjusted to maximize the lower-bound free energy.

The potential $V(\sigma)$ in (2.4) must be chosen so that the sum over σ configurations can actually be evaluated and so that $\langle V \rangle_H = 0$. Both of these conditions may be met with bond-moving potentials, which shift interactions about on the lattice so as to decouple some of the spins. Suppose that $H(\sigma)$ is a sum of nearest-neighbor Ising interactions of the form $K\sigma_i\sigma_j$. An example of a bond-moving potential is $V(\sigma) = K(\sigma_3\sigma_4 - \sigma_1\sigma_2)$. Adding $V(\sigma)$ to $H(\sigma)$ shifts the bond between spins 1 and 2 to spins 3 and 4, as indicated in Fig.2.1. $\langle V \rangle_H$ is zero because of translational invariance. If the system with Hamiltonian H is translationally invariant, $\langle V \rangle_H$ clearly vanishes for any V that adds arbitrary interactions somewhere on the lattice and subtracts them elsewhere.

The energy of a state with all spins parallel is unchanged by bond moving. Consequently, in the case of systems with ferromagnetic ground states, transformations in which the only approximation is bond moving give the free energy exactly in the zero-temperature limit.

---

1 In this chapter a minus sign is incorporated in the definition of the free energy, following the convention in the original papers [2.2,3] on the variational approach. In Chap.1 the minus sign is omitted. Note that the terms upper and lower bound used here are interchanged if the second convention is followed.

2.2.2 Upper-Bound Property of the First-Order Cumulant Approximation

The upper-bound transformation which has received the most attention [2.2,3,36,37] is the first-order cumulant approximation [2.1]. In this approximation H( ) in (2.2) is written as $H = H_0 + H_1$, and H_1 is treated to first order in perturbation theory. The upper-bound property may be derived as follows. Introducing the abbreviations $Z_0(\sigma') = \sum_\sigma P(\sigma',\sigma)\exp[H_0(\sigma)]$ and $\langle A\rangle_0' = Z_0(\sigma')^{-1}\sum_\sigma P(\sigma',\sigma)A(\sigma)\exp[H_0(\sigma)]$, one may write the exact renormalization prescription (2.2) and the first-order cumulant approximation [2.1] in the forms

$$\exp H'_{\text{exact}}(\sigma') = Z_0(\sigma')\,\langle\exp H_1\rangle_0' \quad , \tag{2.6}$$

$$\exp H'_{\text{approx}}(\sigma') = Z_0(\sigma')\,\exp\langle H_1\rangle_0' \quad , \tag{2.7}$$

respectively. The inequality $\exp x \geqslant 1 + x$ implies that $\langle\exp H_1\rangle_0' = \langle\exp(H_1 - \langle H_1\rangle_0')\rangle_0'$ $\exp\langle H_1\rangle_0' \geqslant \exp\langle H_1\rangle_0'$. Combining this result with (2.6,7), one sees that $\exp H'_{\text{exact}}(\sigma') \geqslant \exp H'_{\text{approx}}(\sigma')$, i.e., the Boltzmann factor in the first-order cumulant transformation gives a lower bound to the exact Boltzmann factor for arbitrary configurations of the σ' spins. An infinite sequence of first-order cumulant transformations clearly generates an upper bound to the exact free energy $F = -\ln(Z)$ corresponding to the initial Hamiltonian.

2.3 The Migdal-Kadanoff Transformation

2.3.1 Application to the Ising Model with Nearest-Neighbor Interactions

The Migdal-Kadanoff renormalization transformation was invented by MIGDAL [2.33, 34] and reinterpreted using bond-moving ideas by KADANOFF [2.35]. In KADANOFF's derivation, bonds are moved so that spins may be eliminated by a one-dimensional decimation [2.38]. In the decimation, the sum over a fraction of the spin variables is evaluated in the expression for the partition function, leaving a spin system with a lattice constant that has been enlarged by a factor b and with renormalized interactions. In Fig.2.2, which corresponds to b = 3, the dots denote the decimated spins and the crosses the surviving spins. If the original system only has nearest-neighbor interactions, the renormalized interactions also only involve nearest neighbors.

For an Ising model with nearest-neighbor interactions K, the renormalized coupling K' resulting from a one-dimensional decimation with an integer scale factor b may be calculated explicitly by evaluating the sum

$$\exp(K_0'+K'\sigma_A'\sigma_B') = \sum_{\sigma_1,\dots,\sigma_{b-1}} \exp[K(\sigma_A'\sigma_1+\sigma_1\sigma_2+\dots+\sigma_{b-2}\sigma_{b-1}+\sigma_{b-1}\sigma_B')] \tag{2.8}$$

in which the spins $\sigma_1,\dots,\sigma_{b-1}$ separating two surviving spins σ_A', σ_B' are decimated. Transforming each of the exponential factors in (2.8) according to

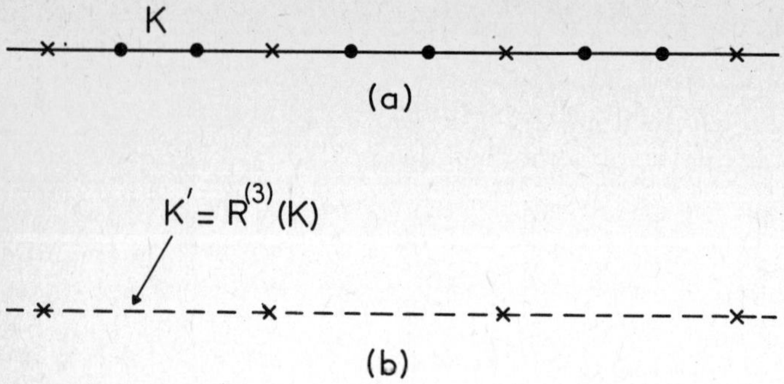

Fig.2.2a,b. One-dimensional system of spins before and after decimation of the spins denoted by dots. The crosses indicate spins that are left after the decimation

$\exp(K\sigma_i\sigma_j) = (\cosh K)(1 + \sigma_i\sigma_j \tanh K)$ and carrying out the sum, one finds the recurrence relation $\tanh K' = (\tanh K)^b$ or $K' = R^{(b)}(K)$, where

$$R^{(b)}(K) = \tanh^{-1}[(\tanh K)^b] \quad . \tag{2.9}$$

This recurrence relation may be readily continued to noninteger values of b.

MIGDAL's recurrence relation [2.33] for the Ising model on a d-dimensional hypercubic lattice is a simple generalization of the one- dimensional result. It has the form

$$K' = R^{(b)}(b^{d-1}K) \quad , \tag{2.10}$$

which is exact for d = 1. Equation (2.10) may be readily derived by dividing the system into hypercubes of b^d spins, moving all the bonds onto the edges of the hypercubes, and then decimating all except the corner spins, as shown for d = 2, b = 3 in Fig.2.3.

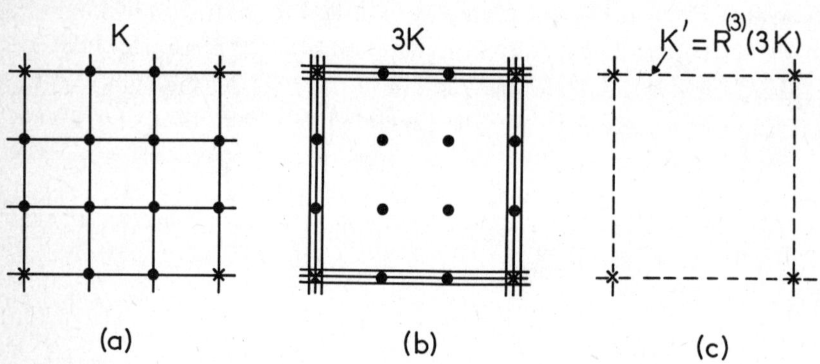

Fig.2.3a-c. Bond-moving and decimation operations corresponding to (2.10) with d = 2, b = 3

KADANOFF [2.35] has proposed a modified version of Migdal's renormalization transformation for anisotropic systems. He considers a d-dimensional hypercubic lattice with distinct couplings $K_1, K_2, \ldots, K_d$ parallel to the 1st, 2nd,...,dth

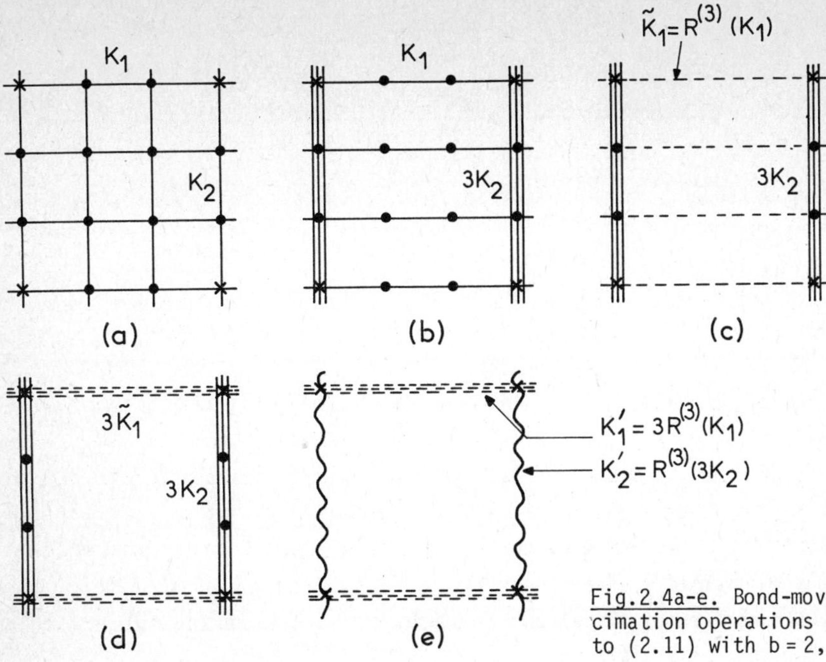

Fig.2.4a-e. Bond-moving and decimation operations corresponding to (2.11) with b = 2, d = 3

axes. In each renormalization step, first all bonds except K_1 are moved, as shown in Fig.2.4, so that a one-dimensional decimation only involving K_1 can be performed. These operations yield new effective couplings $\tilde{K}_1 = R^{(b)}(K_1)$, $\tilde{K}_2 = bK_2, \ldots, \tilde{K}_d = bK_d$. Next all bonds except $\tilde{K}_2$ are moved, and a decimation only involving $\tilde{K}_2$ is performed. After a complete cycle of such operations in which all d axes are considered, renormalized couplings given by

$$K'_\alpha = b^{d-\alpha} R^{(b)}(b^{\alpha-1}K_\alpha) \tag{2.11}$$

are obtained, where $\alpha = 1,2,\ldots,d$. Equation (2.10) becomes identical with (2.11) if K and K' in (2.10) are replaced by $b^{\alpha-d} K_\alpha$ and $b^{\alpha-d} K'_\alpha$, respectively. Thus (2.11) has a fixed point $K^*_\alpha = b^{d-\alpha} K^*$, $\alpha = 1,2,\ldots,d$, for each fixed point K^* of (2.10). Note that the fixed point of (2.11) is anisotropic except in the limit $b \to 1$ of an infinitesimal transformation. This is a convenient limit if the initial Hamiltonian is isotropic, since the transformation (2.11) otherwise generates an unnatural anisotropy.

The differential equation to which the infinitesimal transformation corresponds is given by

$$\frac{dK}{d\ell} = (d-1)K + (\sinh K)(\cosh K)\ln\tanh K. \tag{2.12}$$

It follows from (2.9) and either (2.10) or (2.11) on setting $b = 1 + \delta\ell$ and keeping first-order terms in the infinitesimal quantity $\delta\ell$.

In cases where the initial Hamiltonian is isotropic and the decimation sufficiently complicated that one is forced to use numerical methods with integer values of b, MIGDAL's isotropic recursion relation (2.10) conveniently retains the isotropy, whereas KADANOFF's anisotropic formula (2.11) does not. A nice property of the anisotropic transformation is that it preserves the dual symmetry [2.35] of the anisotropic two-dimensional q-state Potts model, of which the Ising model is a special case, and thus gives the critical line exactly. For example, for d = 2 and a given value of the scale factor b, (2.11) with $R^{(b)}(K)$ defined by (2.9) has a single fixed point with K_1^* and K_2^* both finite and nonzero. K_1^* and K_2^* satisfy $K_1^* = bK_2^*$ and the exact criticality condition $\sinh 2K_1^* \sinh 2K_2^* = 1$. In the limit $b \to 1$, which corresponds to the infinitesimal transformation (2.12), the fixed point $K_1^* = K_2^* = (1/2)\ln(1 + \sqrt{2})$ is identical with the exact critical coupling of the isotropic d = 2 Ising model. One would expect an exact renormalization transformation to reproduce the entire critical line for a given value of b, not just a single point on the critical line. However, this cannot be achieved in an approximation in which K_1 and K_2 transform independently.[2]

The Migdal-Kadanoff renormalization procedure is much less successful in predicting the critical exponents of the d = 2 Ising model than the critical coupling. A value for the thermal scaling index $y_t = \nu^{-1}$ may be extracted from the largest eigenvalue of the renormalization transformation linearized about the fixed point [2.1]. Formulae (2.10,11) both yield the same eigenvalue.[3] For $b \to 1$ one [2.33] obtains $y_t = 0.754$ as compared with the exact value $y_t = 1$. The discrepancy increases [2.35] monotonically with b. The lower critical dimensionality of the Ising model, at which the critical temperature goes to zero, is $d_{lc} = 1$. As discussed in Sect.2.2, in ferromagnetic systems the bond-moving prediction is exact in the limit $T \to 0$. MIGDAL [2.33] speculates that in dimension $d = 1 + \varepsilon$, the first order predictions $y_t = \varepsilon$, $(K^*)^{-1} = 2\varepsilon$, which are obtained for any choice of the scale factor b, are exact. The result for y_t has been confirmed by a field-theoretical approach [2.39, 40] in $1 + \varepsilon$ dimensions. MIGDAL also considers the n-vector model with $n > 2$, for which $d_{lc} = 2$, in $2 + \varepsilon$ dimensions. The first-order prediction for y_t also agrees with the field-theoretical results [2.41,42]. The critical temperature he finds is proportional to ε but with a numerical coefficient different from that predicted by the field theory.

As mentioned above, the bond-moving approximation becomes exact in the zero-temperature limit. The Migdal-Kadanoff renormalization, which yields a lower bound to the true free energy, reproduces both the low- and high-temperature limits

2 R.K.P. ZIA and R. LIPOWSKY (unpublished) have found that the flow properties can be significantly improved by rotating as well as translating bonds to couple the different K_α.

3 Note that (2.11) predicts an unphysical d-fold degeneracy of the eigenvalue.

$f(K \to \infty) \to -Kd$, $f(K \to 0) \to -\ln 2 - (d/2)K^2$ exactly [2.43]. For a comparison of the exact and lower-bound free energies for arbitrary values of K, the reader is referred to [2.43].

2.3.2 Inclusion of a Magnetic Field

In order to calculate [2.1] the magnetic scaling index $y_h = d\delta/(1+\delta)$, the Migdal-Kadanoff transformation must be generalized to include a magnetic field. Adding interactions of the form $h\sigma_i$ to $H(\sigma)$ and decimating the one-dimensional Ising chain, one finds the transformation $h' = \Lambda^{(b)}(K)h + O(h^3)$, where

$$\Lambda^{(b)}(K) = \frac{1+\tanh K}{1-\tanh K}\,\frac{1-(\tanh K)^b}{1+(\tanh K)^b} \quad . \tag{2.13}$$

The extent to which the single-spin interactions $h\sigma_i$ should be moved in higher dimensions before decimating is ambiguous. If the single-spin interactions are not moved at all, and the nearest-neighbor interactions are redistributed as in the derivation of (2.10) (Fig.2.3), one obtains the transformation

$$h' = \{1 + d[\Lambda^{(b)}(b^{d-1}K) - 1]\}h + O(h^3) \quad . \tag{2.14}$$

MIGDAL [2.33] proposed an alternate formula, which may be derived by associating a combined interaction $K\sigma_i\sigma_j + (h/2d)(\sigma_i + \sigma_j)$ with each bond and moving both the single-spin and two-spin terms. This procedure, in which the spins in Fig.2.3 that are completely decoupled from their nearest neighbors are also decoupled from the magnetic field, corresponds to the transformation

$$h' = b^{d-1}\Lambda^{(b)}(b^{d-1}K)h + O(h^3) \quad . \tag{2.15}$$

The bond-moving approximation corresponding to (2.14) is exact in the limit $K = 0$ and for finite nonzero K can be shown to yield a lower bound to the exact free energy as high or higher than the bond-moving procedure leading to (2.15). In some cases (2.15) gives better predictions for the critical exponents. The approximations yielding (2.15) give y_h exactly [2.41,42] to first order in ε in the case of the n-vector model with $n > 2$ in $2 + \varepsilon$ dimensions. For the two-dimensional Ising model (2.15) yields $y_h = 1.881$ in the limit $b \to 1$, as compared with the exact value 15/8, whereas (2.14) gives $y_h = 1.763$. MIGDAL suggested that the first-order prediction $y_h = 1 + \varepsilon$ for the Ising model in $1 + \varepsilon$ dimensions, which is obtained for arbitrary b with (2.15) and in the limit $b \to 1$ with (2.14), may be exact. Note that for arbitrary b (2.15) correctly reproduces the exact [2.1,44] value $y_h = d$ for the low-temperature discontinuity fixed point $K^* = \infty$, whereas (2.14) only predicts the exact value in the limit $b \to 1$.

2.3.3 The Bond-Moving Prescription of EMERY and SWENDSEN

The ambiguity in moving single-spin interactions also arises in models with crystal-field interactions, such as the Blume-Emery-Griffiths model [2.45], which has the spin-one Hamiltonian:

$$H(s) = \sum_{<ij>} (Js_i s_j + K s_i^2 s_j^2) - \Delta \sum_k s_k^2 \quad , \tag{2.16}$$

where $s_i = \pm 1,0$. If the prescription of MIGDAL leading to (2.15) is followed, one moves the combined bond $Js_i s_j + K s_i^2 s_j^2 - (\Delta/q)(s_i^2 + s_j^2)$, where q is the coordination number of the lattice, as in Fig.2.3. In the case of configurations with all spins up or down, the combined bond is a large quantity at low temperatures, where J, K, and Δ are large, unless $J + K - 2\Delta/q$ is small. EMERY and SWENDSEN [2.46] have recommended an alternate bond-moving procedure. They write the Hamiltonian in the form

$$H(s) = \sum_{<ij>} \hat{J}(s_i,s_j) + \sum_k G(s_k^2) \tag{2.17}$$

and choose $\hat{J}(s_i,s_j)$ so that it vanishes if $s_i = s_j$. Then the two-spin interactions $\hat{J}(s_i,s_j)$ are moved but not the single-spin interactions $G(s_k^2)$. For the Blume-Emery-Griffiths model $\hat{J}(s_i,s_j) = -(1/2)J(s_i - s_j)^2 - (K/2)(s_i^2 - s_j^2)^2$, and $G(s_k^2) = [(q/2)(J + K) - \Delta]s_k^2$. With this decomposition the bond $\hat{J}(s_i,s_j)$ that is moved is a small quantity both at high temperatures and for the dominant configurations with all the s_i equal at low temperatures. The advantages and disadvantages of these two procedures in applications to the Blume-Emery-Griffiths model and the Potts lattice gas are discussed in [2.17c,46-48].

2.3.4 Inconsistent Scaling of the Correlation Function

Since the spins are eliminated by decimating, the Migdal-Kadanoff transformation suffers from a well-known inconsistency in the scaling of the spin-spin correlation function [2.1,38]. In the decimation transformation, the correlation function $g_H(r) = <\sigma(r)\sigma\,0>_H$, with r measured in units of the lattice constant, transforms according to $g_H(r) = g_{H'}(r/b)$. This transformation is inconsistent with the expected algebraic decay $g_{H^*}(r) \sim r^{-(d-2+\eta)}$ of the correlation function at the fixed point in systems where $d - 2 + \eta = 2\beta/\nu \neq 0$. The unimpressive accuracy with which the Migdal-Kadanoff transformation predicts the exponent ν for the d = 2 Ising model may be related to this inconsistency. Near the lower critical dimensionalities of the Ising and n-vector models with n > 2, where β/ν approaches zero, the transformation predicts several scaling indices exactly, as mentioned above.

2.3.5 Relation to Exactly Soluble Hierarchical Models

FORGACS and ZAWADOWSKI [2.49] and BERKER and OSTLUND [2.50] have pointed out that the Migdal-Kadanoff renormalization procedure is exact for a class of models on hierarchical lattices that lack translational invariance. KAUFMAN and GRIFFITHS [2.51] have studied some of the unusual properties (continuously varying critical exponents, phase transitions without long-range order) of these systems. Figure 2.5 indicates how hierarchical lattices for which the transformations $K' = R^{(2)}(2K)$ and $K' = 2R^{(2)}(K)$ are exact may be generated.

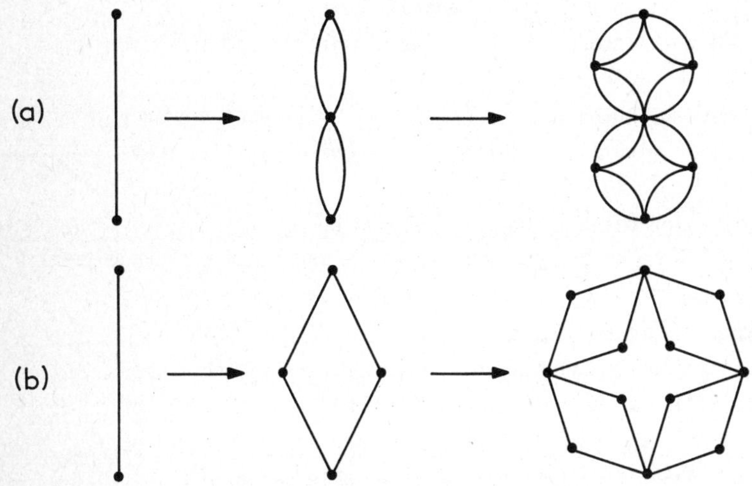

Fig.2.5a,b. Generation of hierarchical lattices for which the transformations $K' = R^{(2)}(2K)$ and $K' = 2R^{(2)}(K)$, respectively, are exact. The lattices are produced by replacing a single bond by 4 bonds as shown, making the same replacement in the resulting structure, and continuing this process

2.3.6 Applications

Because of its mathematical simplicity the Migdal-Kadanoff renormalization procedure has been applied to a wide variety of different systems. One hopes to obtain fairly reliable predictions for critical exponents near the lower critical dimension and for the phase diagrams of Ising-like systems near two dimensions. In addition to the applications mentioned above, the x-y model [2.52], the x-y lattice gas [2.53,54], the Blume-Emery-Griffiths model [2.46,47], the Potts lattice gas [2.17c,48], the roughening transition [2.55], models of adsorbed monolayers [2.50, 56-58], Ising models with Lifschitz points [2.59], the semiinfinite Ising model [2.43], disordered Ising models [2.60-64], Ising models on fractal lattices [2.65], the kinetic Ising model [2.66-68], quantum spin systems [2.21,69-71], and lattice gauge theories [2.34,35,72,73] have been considered. This list of applications is far from complete, and I apologize to those whose work has not been mentioned.

2.3.7 Modifications of the Migdal-Kadanoff Procedure

In the case of a spin system on a square lattice with only nearest-neighbor interactions, an exact decimation transformation which eliminates half of the spins and which, in general, produces nearest-neighbor, next-nearest-neighbor, and four-spin interactions can be carried out [2.38]. There have been attempts to improve the Migdal-Kadanoff transformation for two-dimensional systems by combining this type of decimation [2.74] or a duality-decimation relation [2.75] with approximations in which none of the nearest-neighbor interactions are moved. These procedures, which are difficult to extend to other dimensions, give considerably better values for the scaling index $y_t = \nu^{-1}$ than the Migdal-Kadanoff method. The calculation in [2.74] suffers from the same inconsistency in the scaling of the spin-spin correlation function.

In designing bond-moving approximations for systems with nonferromagnetic ground states, one should keep in mind that arbitrary bond moving does not leave the ground-state energy invariant, as in the ferromagnetic case. In [2.76] an approximate renormalization transformation based on bond moving and decimation is used to calculate the global phase diagram of the two-dimensional Ising model on a square lattice with nearest-neighbor, next-nearest-neighbor, and four-spin interactions, which has several different types of ground states.

MARTINELLI and PARISI [2.77] have attempted to improve the Migdal-Kadanoff renormalization transformation with a systematic perturbation theory which is applicable, in principle, to all approximate transformations of the form (2.4). They replace $V(\sigma)$ in (2.4) by $(1 - \varepsilon)V(\sigma)$, where ε is a variable parameter, thereby obtaining a transformation of the form $H' = R[H + (1 - \varepsilon)V]$. The case $\varepsilon = 0$ corresponds to a simple zeroth-order approximation $H' = R[H + V]$ such as the Migdal-Kadanoff formula, and the case $\varepsilon = 1$ to the exact renormalization transformation $H' = R[H]$. In [2.77,78] critical exponents for $\varepsilon = 1$ are estimated from the leading terms in a perturbation series in ε.

2.4 Variational Transformations

2.4.1 The Kadanoff Lower-Bound Variational Transformation

This section begins with a discussion of KADANOFF's one-hypercube lower-bound variational transformation [2.2,3], the most widely used variational transformation. As in the previous section, the Ising model is considered as a specific example.

KADANOFF's transformation is applicable to d-dimensional spin systems on a hypercubic lattice with Hamiltonians [see (2.4)] of the form $H = \sum\hat{H}$, where the sum goes over all elementary hypercubes of the lattice. Since each elementary hypercube has 2^d sites, $\hat{H}$ depends on 2^d spin variables σ. Each σ spin occurs in 2^d adjacent hypercubes. In a fraction 2^{-d} of the elementary hypercubes, renormalized spin variables σ', denoted by crosses in Fig.2.6, are introduced. The σ' spins are situated on a hypercubic lattice with twice the lattice constant of the lattice of σ spins.

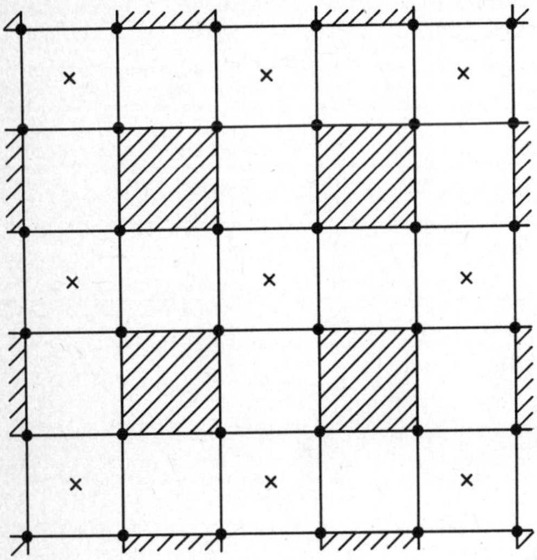

Fig.2.6. Application of the Kadanoff-lower-bound variational transformation to the Ising model on a square lattice. The crosses indicate σ' spins assigned to cells of 4 σ spins. The bond-moving operation shifts σ interactions in the Hamiltonian and in the weight function into the shaded squares

A weight function is chosen that couples each σ' spin to the spins $\sigma_1,\ldots,\sigma_{2^d}$ of the hypercube according to the formula

$$\hat{P}(\sigma';\sigma_1,\ldots,\sigma_{2^d}) = \frac{\exp[p\sigma'(\sigma_1 + \ldots + \sigma_{2^d})]}{2\cosh[p(\sigma_1 + \ldots + \sigma_{2^d})]} \quad . \tag{2.18}$$

The function $P(\sigma',\sigma)$ in (2.2) is given by $P = \Pi'\hat{P}$, where there is a factor $\hat{P}$ for each σ' spin. The requirement (2.3) is satisfied for arbitrary values of the parameter p, which will turn out to be a variational parameter. Note that in the limit $p \to \infty$ the weight function (2.18) corresponds to the "majority rule" of NIEMEIJER and VAN LEEUWEN [2.1], according to which the cell spin has the same sign as the majority of the site spins.

As discussed in Sects.2.1,2, the sum over σ configurations in the exact renormalization transformation (2.2) cannot, in general, be evaluated exactly. KADANOFF's lower-bound transformation replaces H on the right side of (2.2) by H + V, to obtain an approximate renormalization transformation of the form (2.4). The potential V is chosen to move bonds in such a way that the sum in (2.4) can be evaluated explicitly. As explained in Sect.2.2, this approximation scheme yields a lower bound to the exact free energy. The parameter p in the weight function can be adjusted to maximize the lower bound.

Each elementary hypercube of 2^d σ' spins surrounds a "shaded hypercube" of 2^d σ spins, as shown for d = 2 in Fig.2.6. The bond-moving approximation used by KADANOFF shifts all the interactions in H and all the quantities $u(\sigma_1,\ldots,\sigma_{2^d}) = -\ln\{2\cosh[p(\sigma_1 + \ldots + \sigma_{2^d})]\}$ from the weight function into the shaded hypercubes. The $V(\sigma)$ in (2.4) that accomplishes this has the explicit form

$$V(\sigma) = \sum_{\substack{\text{shaded}\\ \text{hypercubes}}} (2^d\hat{H} + u) - H - \sum_{\substack{\text{hypercubes}\\ \text{with crosses}}} u \quad . \tag{2.19}$$

After the bond moving, each σ spin is only coupled to the other σ spins in the same shaded hypercube and to the nearest σ' spin. The sum on the right-hand side of (2.4) can be factored into sums of the form

$$\exp[\hat{H}'(\sigma_1', \ldots, \sigma_{2^d}')] = \sum_{\sigma_1,\ldots,\sigma_{2^d}} \frac{\exp[p(\sigma_1'\sigma_1+\ldots+\sigma_{2^d}'\sigma_{2^d}) + 2^d H(\sigma_1,\ldots,\sigma_{2^d})]}{2\cosh[p(\sigma_1+\ldots+\sigma_{2^d})]} \tag{2.20}$$

for each shaded hypercube. H' in (2.4) is given by the sum $H' = \sum' \hat{H}'$ over all elementary hypercubes of the lattice of σ' spins.

It is useful to expand $\hat{H}$ and $\hat{H}'$ in the form $\hat{H} = \sum_\alpha K_\alpha S_\alpha$, where the S_α form a complete set of interactions for the 2^d spins in a hypercube that reflects the symmetries of the system of interest, and the K_α are the corresponding coupling constants. For the two-dimensional isotropic Ising model on a square lattice with a Hamiltonian that is invariant under change of sign of all the spins, a convenient complete set consists of the interactions [2.2,3,79]

$$\begin{aligned} S_0 &= 1 \quad , \\ S_{nn} &= \sigma_1\sigma_2 + \sigma_2\sigma_3 + \sigma_3\sigma_4 + \sigma_4\sigma_1 \quad , \\ S_{nnn} &= \sigma_1\sigma_3 + \sigma_2\sigma_4 \quad , \\ S_4 &= \sigma_1\sigma_2\sigma_3\sigma_4 \quad . \end{aligned} \tag{2.21}$$

The subscripts nn and nnn stand for nearest neighbors and next-nearest neighbors, respectively. If the initial Hamiltonian contains interactions that break the up-down spin symmetry, the odd operators

$$\begin{aligned} S_1 &= \sigma_1 + \sigma_2 + \sigma_3 + \sigma_4 \quad , \\ S_3 &= \sigma_1\sigma_2\sigma_3 + \sigma_2\sigma_3\sigma_4 + \sigma_3\sigma_4\sigma_1 + \sigma_4\sigma_1\sigma_2 \quad , \end{aligned} \tag{2.22}$$

must be included as well. It is straightforward to obtain recurrence relations for the coupling constants K_α from (2.20). Some details are given in [2.3,79].

The recurrence relations for the K_α have the structure

$$K_0' = b^d[K_0 + g(\underline{K},p)] \quad , \tag{2.23}$$

$$\underline{K}' = \underline{K}'(\underline{K},p) \quad . \tag{2.24}$$

Here $\underline{K}$ denotes all of the coupling constants except K_0, the constant associated with $S_0 = 1$ in (2.21). Greek indices will be used below for the entire set $(K_0,\underline{K})$ and latin indices for the $\underline{K}$. The first term on the right side of (2.23) represents the contribution to the new constant term from the previous constant term, and the second term is the contribution from the interacting part of the Hamiltonian. The quantity $b^d = 2^d$ is the number of site spins per cell spin. Since the constant term does not lead to interactions on renormalizing, the $\underline{K}$ transform independently of K_0, as indicated in (2.24).

Equations (2.4,23,24) imply a recurrence relation for the lower-bound free energy per spin $f = -\lim N^{-1} \ln Z$, $N \to \infty$. For an initial Hamiltonian with couplings $K_0 = 0$, $\underline{K}$, and a sequence of variational parameters $p,p',p'',\ldots$ in successive renormalization transformations, one finds

$$f(\underline{K};p,p',\ldots) = -g(\underline{K},p) + b^{-d} f(\underline{K}'(\underline{K},p);p',p'',\ldots) \quad . \tag{2.25}$$

2.4.2 The Kadanoff Criterion for the Optimal Variational Parameter

In computing critical exponents, KADANOFF [2.2,3] uses the same value of p in each application of the renormalization transformation rather than a sequence of different values, so that (2.25) is replaced by

$$f(\underline{K},p) = -g(\underline{K},p) + b^{-d} f(\underline{K}'(\underline{K},p),p) \quad . \tag{2.26}$$

The location of the fixed point $\underline{K}^*(p)$ of (2.24) and the values of the critical exponents depend, of course, on p. The optimal p chosen according to the Kadanoff criterion is the value p^* that maximizes $f(\underline{K},p)$ at the fixed point, i.e., the value for which $\partial f(\underline{K},p)/\partial p$ vanishes at $\underline{K} = \underline{K}^*(p)$.

The Kadanoff criterion can be put in a more convenient form for calculations. Denoting the derivatives of $f(\underline{K},p)$ with respect to K_i and p, evaluated at the fixed point, by $\partial f^*/\partial K_i$ and $\partial f^*/\partial p$ and differentiating (2.26), one finds

$$\frac{\partial f^*}{\partial K_i} = -\frac{\partial g^*}{\partial K_i} + \frac{1}{b^d}\sum_{j=1}^{m} \frac{\partial f^*}{\partial K_j}\frac{\partial K'_j}{\partial K_i} \quad , \tag{2.27}$$

$$(1 - b^{-d})\frac{\partial f^*}{\partial p} = -\frac{\partial g^*}{\partial p} + \frac{1}{b^d}\sum_{j=1}^{m} \frac{\partial f^*}{\partial K_j}\frac{\partial K'_j}{\partial p} \quad . \tag{2.28}$$

The upper summation index m is the number of couplings in $\underline{K}$. Introducing $\psi_0 = 1$, $\psi_i = -\partial f^*/\partial K_i$ and recalling (2.23), one may write (2.27,28) in the forms

$$\sum_{\alpha=0}^{m} \psi_\alpha \frac{\partial K'_\alpha}{\partial K_\beta} = b^d \psi_\beta \quad , \quad \beta = 0,1,2,\ldots,m \quad . \tag{2.29}$$

$$\frac{\partial f^*}{\partial p} = -\frac{1}{b^d - 1}\sum_{\alpha=0}^{m} \psi_\alpha \frac{\partial K'_\alpha}{\partial p} \quad . \tag{2.30}$$

From (2.29) one sees that ψ_α is a left eigenvector with eigenvalue b^d of the matrix[4] of derivatives $\partial K'_\alpha/\partial K_\beta$ evaluated at the fixed point. According to (2.30) the inner product of this eigenvector with $\partial K'_\alpha/\partial p$ vanishes for the value of p that satisfies[5] the Kadanoff criterion $\partial f^*/\partial p = 0$. In practice it is particularly convenient to apply the Kadanoff criterion by working with (2.29,30). Once the inner product in (2.30) vanishes, predictions for critical indices and the corresponding eigenoperators can be obtained from the other eigenvalues and eigenvectors [2.1] of the matrix $\partial K'_\alpha/\partial K_\beta$.

The lower-bound variational transformation with the variational parameter fixed at the value for which $\partial f^*/\partial p$ vanishes predicts critical exponents for a variety of models with an impressive accuracy that has not been achieved by any other real-space renormalization methods except phenomenological renormalization (Chap.1), a finite-size scaling technique that is considerably more laborious. Table 2.1 compares predictions for the critical exponents of the spin-1/2 Ising model in d = 2,3,4 dimensions [2.2,3,80] and for the d = 2 tricritical exponents of the Blume-Emery-Griffiths model [2.10,11] with exact values or best estimates [2.17,18,81-84]. A list of other applications of the lower-bound variational method is given below.

Table 2.1. Predictions of the Kadanoff lower-bound variational transformation for the Ising critical exponents in d = 2,3,4 dimensions [2.2,3,80] and the Blume-Emery-Griffiths tricritical exponents in d = 2 dimensions [2.10,11]. The numbers in parentheses are exact results, best estimates, or likely conjectures.

Critical exponents	d = 2	d = 3	d = 4
$2 - \alpha = d\nu$	1.998 (2)[a]	1.887 (1.890 ± 0.0024)[b]	1.965 (2)[c]
δ	15.04 (15)[a]	4.604 (4.82 ± 0.03)[b]	2.90 (3)[c]
Tricritical exponents	d = 2		
$2 - \alpha_t = d\nu_t$	1.113 (10/9)[d]		
δ_t	26.59 (77/3)[e]		
ϕ_t	0.4445 (4/9)[e]		

[a]Exact result; [b]best estimate [2.81]; [c]exact (mean-field) result; [d]exact result [2.17,82,83]; [e]conjecture [2.17,18,84]

4 One can prove [2.79] that the matrix, which is real and nonsymmetric, only has real eigenvalues, so that ψ_α can be chosen to be real. From (2.23,24) it follows that the right eigenvector ϕ_α of the matrix with eigenvalue b^d is given by $\phi_\alpha = (\phi_0,0,0,\ldots,0)$.

5 One should verify that p^* corresponds to a maximum, not just an extremum. A puzzling situation in which a $\underline{p}^*$ that corresponds to a saddle point rather than a maximum gives excellent predictions for critical exponents is described in [2.17,18].

2.4.3 Problems with the Lower-Bound Variational Transformation

Despite the success in predicting critical exponents, there are certain fundamental difficulties with the lower-bound transformation. From (2.20) one sees that the transformation has an invariant subspace in which $\hat{H}(\sigma_1,\ldots,\sigma_{2d})$ is symmetric in $\sigma_1,\ldots,\sigma_{2d}$, i.e., is invariant under interchange of any two of the σ_i. The predictions in Table 2.1 are obtained for fixed points in this symmetric subspace. For the Ising model in d = 2,3 dimensions, there is a spurious weakly-relevant eigenvector [2.79,85] at this fixed point that points outside the symmetric subspace. An initial Hamiltonian with only nearest-neighbor interactions is not in the symmetric subspace[6] and under application of the lower-bound transformation is mapped onto a second fixed point. This fixed point [2.79,85] also yields excellent values for the d = 2 Ising exponents when p is optimized according to the Kadanoff criterion, though less impressive than the values for the symmetric fixed point. Thus the Kadanoff transformation has two fixed points where there should be only one.

KADANOFF et al. describe some other disturbing aspects of the lower-bound variational transformation in their original work [2.3] on the method. The transformation is less successful in predicting the critical temperature and other nonuniversal quantities than the critical exponents. The impressive predictions of critical exponents are obtained with the parameter in the renormalization transformation fixed at the optimal value at the fixed point, whereas a higher lower-bound to the exact free energy is obtained if p,p',p'',... in (2.25) are regarded as independent variational parameters. The many-parameter optimization leads to a renormalization transformation of the form

$$\underline{K}' = \underline{K}'(\underline{K},p(\underline{K})) \quad , \tag{2.32}$$

where $p(\underline{K})$ is the optimal value of p for the renormalization step beginning with $\underline{K}$. KNOPS [2.85] pointed out that the $\underline{K}$ dependence of $p(\underline{K})$ should also be taken into account in linearizing the renormalization transformation about the fixed point and estimated that the extra $\underline{K}$ dependence considerably worsens the predictions of the critical index $y_t = \nu^{-1}$.

---

6 The best predictions of critical exponents are generally obtained in the symmetric subspace, where the transformation is especially simple, since the number of coupling constants that must be considered is small. An initial Hamiltonian with nearest-neighbor and single-spin interactions on a body-centered hypercubic lattice may be mapped exactly into the symmetric subspace by decimating one of the two simple hypercubic sublattices of which the body-centered hypercubic lattice is composed. In the case of the Ising model with nearest-neighbor interactions only, the decimation generates a symmetric $\hat{H}$ given by

$$\begin{aligned}\exp[\hat{H}(\sigma_1,\ldots,\sigma_{2d})] &= \sum_{\sigma_0} \exp[K(\sigma_1+\ldots+\sigma_{2d})\sigma_0] \\ &= 2\cosh[K(\sigma_1+\ldots+\sigma_{2d})] \quad . \end{aligned} \tag{2.31}$$

The behavior of the function $p(\underline{K})$ in the neighborhood of the fixed point has been studied in detail by VAN SAARLOOS et al. [2.4] and by DEN NIJS and KNOPS [2.5]. The result, that in general $p(\underline{K})$ is singular at $\underline{K}$, has rather serious implications for the variational approach to real-space renormalization, which are discussed in the next section.

2.4.4 Determination of an Optimal Sequence of Variational Parameters

The structure of the equations determining $p(\underline{K})$, the optimal variational parameter in the renormalization step beginning with $\underline{K}$, and the conclusions of VAN SAARLOOS et al. [2.4] and DEN NIJS and KNOPS [2.5] concerning singularities in $p(\underline{K})$ will now be outlined briefly. Note that the free energy f on the right side of (2.25) only depends on p through the coupling constants $\underline{K}'$ [2.4-6]. Denoting the free energy optimized with respect to p,p',p",... by $f(\underline{K})$, one finds from (2.25) and its derivative with respect to p

$$f(\underline{K}) = -\,g(\underline{K},p(\underline{K})) + \frac{1}{b^d}\,f(\underline{K}') \quad , \tag{2.33}$$

$$0 = -\frac{\partial g(\underline{K},p(\underline{K}))}{\partial p} + \frac{1}{b^d}\sum_{i=1}^{m}\frac{\partial f(\underline{K}')}{\partial K_i'}\,\frac{\partial K_i'(\underline{K},p(\underline{K}))}{\partial p} \quad ; \tag{2.34}$$

$f(\underline{K})$ and $p(\underline{K})$ are determined by solving (2.32-34) simultaneously.

In some special cases $\partial g(\underline{K},p)/\partial p$ and the $\partial K_i'(\underline{K},p)/\partial p$ all vanish for the same $p(\underline{K})$. Then (2.34) is satisfied trivially, and $p(\underline{K})$ is uncoupled from any singularities in $f(\underline{K})$. This is the case for the Kadanoff lower-bound variational transformation applied to the one-dimensional Ising model. Both derivatives vanish and the transformation (2.20) becomes identical with the exact one-dimensional decimation (2.8) with $b = 2$ on choosing $\cosh[2p(K)] = \exp(2K)$. The first-order cumulant expansion was shown in Sect.2.2 to be an example of an upper-bound transformation. In this approximation the derivative $\partial g(K,p)/\partial p$ determined with the weight function (2.18) is zero [2.4] for all values of p. The derivative $\partial K'(K,p)/\partial p$ only vanishes for $p = \infty$. In this limit the weight function reduces to the majority-rule weight function of NIEMEIJER and VAN LEEUWEN [2.1]. BARBER has shown more generally [2.36] that the optimal upper-bound free energy in the first-order cumulant approximation is obtained with the cell spins chosen according to the majority rule.

Let us now turn to the nontrivial case in which the different terms in (2.34) do not all vanish simultaneously for the optimal $p(\underline{K})$ and in which the optimal $p(\underline{K})$ does not correspond to some sort of boundary extremum such as $p(\underline{K}) = \infty$. Then, in general, $p(\underline{K})$ is singular at the fixed point.[7] The singular behavior that follows

7 It is easy to verify that the value $p(\underline{K}^*)$ at the fixed point obtained by solving (2.32-34) is the same as the value p^* determined by the Kadanoff criterion that $\partial f^*/\partial p$ in (2.30) vanish.

from (2.32-34) is analyzed in detail in [2.4,5], and only the results will be recorded here. One finds the following possibilities.

a) $p(\underline{K})$ is a singular function at the fixed point with exponent $1 - \alpha$, $\alpha < 0$. No consistent solution with $\alpha > 0$ is possible.

b) There is a discontinuous transition between different branches of the optimal $p(\underline{K})$.

For completeness the two situations mentioned above that lead to a nonsingular $p(\underline{K})$ will also be listed here.

c) The derivatives $\partial g/\partial p$ and $\partial K_j'/\partial p$ in (2.34) all vanish for the same $p(\underline{K})$, so that $p(\underline{K})$ is uncoupled from the singularity in $f(\underline{K})$.

d) The optimal $p(\underline{K})$ corresponds to the boundary of the allowed domain, for example, $p(\underline{K}) = \infty$.

The optimal $p(\underline{K})$ for KADANOFF's lower-bound variational transformation in the case of the two-dimensional Ising model belongs to category a). The corresponding critical exponent $2 - \alpha = d\nu$ has [2.4,5] the value 2.1234 as compared with the exact result 2 and the value 1.998 originally obtained by KADANOFF [2.2,3] neglecting variations of the variational parameter about its optimal value at the fixed point. Taking the $\underline{K}$ dependence of $p(\underline{K})$ into account improves the free energy but destroys the excellent agreement between the calculated and exact values of $2 - \alpha = d\nu$.

The analysis of VAN SAARLOOS et al. [2.4] and DEN NIJS and KNOPS [2.5] creates serious doubts concerning the usefulness of the variational approach in calculations of critical exponents. Since the variational principle involves the free energy, which is a singular function of $\underline{K}$, the optimal function $p(\underline{K})$ is also, in general, a singular function of $\underline{K}$. Thus a basic assumption of the renormalization-group approach, that the $\underline{K}'$ are nonsingular functions of the $\underline{K}$, is violated. In the nontrivial case in which either a) or b) applies, the method appears incapable of describing a system with positive α. Disregarding the $\underline{K}$ dependence of the variational parameter, as in KADANOFF's original calculations of critical exponents, avoids all these fundamental difficulties and makes the weight function independent of the renormalization step, as in other real-space renormalization methods. Calculating exponents with a fixed value of p is not in itself unreasonable. However, one feels uneasy about using a variational principle to pick the best single value of p, when improving the variational calculation for the free energy by permitting p to depend on $\underline{K}$ leads to the fundamental difficulties mentioned above.

2.4.5 Applications of the Lower-Bound Variational Transformation

KADANOFF's lower-bound variational transformation has been applied to numerous systems besides the Ising model. For most of these systems astonishingly accurate predictions of the critical exponents are also obtained when the variational parameter is fixed at its optimal value at the fixed point, as in KADANOFF's original procedure. For the n-vector model in $4-\varepsilon$ dimensions, the usual thermal and magnetic critical exponents [2.3] and the correction to scaling exponents [2.7] are obtained exactly to first order in ε. With a different criterion for p the exponent ν is also obtained [2.8] exactly to first order in the 1/n expansion in dimensions 2 and 4. The n-vector model with cubic anisotropy is considered in [2.9]. KADANOFF's transformation has also been applied to other systems with discrete spin variables besides the Ising model, for example, the spin-1 Ising or Blume - Emery-Griffiths model [2.10,11], the Potts model [2.5,12-14], and the Ashkin-Teller-Potts model [2.15,16]. The particularly precise results [2.17,18] obtained for the Potts lattice gas confirm the importance of vacancies in the crossover between first- and second-order transitions and lend support to several conjectures for the critical exponents of the two-dimensional model. In [2.19] the discrete Gaussian model is considered and in [2.20] the x-y model. In [2.21] the method is shown to reproduce some exact results for a quantum system, the Luttinger-Thirring model.

Other properties of the Ising model besides the exponents in Table 2.1 have been studied with the lower-bound variational transformation. In [2.3,7,22-24] the free energy and other thermodynamic functions are considered, and in [2.25] the dimensional dependence of the critical exponents. Results for the triangular lattice are reported in [2.26,27] and for the Ising antiferromagnet in a field on a square lattice in [2.28]. In [2.29] the interface free energy of the Ising model is calculated. In [2.30] the Kadanoff transformation is used to calculate surface critical exponents and a phase diagram for the semiinfinite Ising model. The critical temperatures of the spin-s Ising model and a model for structural transitions are calculated in [2.31,32], using lower-bound restructuring transformations to map the models onto the spin-1/2 Ising model in the first renormalization step.

2.4.6 Other Variational Methods

KADANOFF's lower-bound variational transformation and the first-order cumulant approximation, which was shown in Sect.2.2 to be an upper-bound transformation, are the variational transformations that have been studied in most detail. SHENKER et al. [2.37] have considered a modified first-order cumulant approximation in which the perturbation consists of the intracell interactions minus a fraction p of the intercell interactions. The variational principle is used to choose the optimal p at the fixed point. In linearizing the renormalization about the fixed point, SHENKER et al. neglect second- and higher-order terms in the perturbation expansion and set $\partial p/\partial K_i = 0$, which simplifies the calculation by eliminating a second-order cumulant expression. The results for the critical couplings and critical indices of the Ising and Potts models obtained with this procedure are as good or better than the results of the ordinary cumulant approximation.

A variational method different from those mentioned so far has been suggested by BURKHARDT and SOUTHERN [2.74]. Generally, the first renormalization step produces more complicated interactions than are present in the initial Hamiltonian of interest. The next step becomes mathematically tractable if the Hamiltonian is truncated to eliminate the extra interactions. In [2.74] truncation procedures that give upper and lower bounds to the exact free energy are devised. The couplings that are retained in the truncated Hamiltonian play the role of variational parameters. Calculations combining an ordinary decimation with scale factor $b = \sqrt{2}$ and optimal upper- and lower-bound truncations in which only nearest-neighbor couplings are retained give upper and lower bounds to the free energy of the $d = 2$ Ising model that coincide in the high- and low-temperature limits and differ by less than 1% for all temperatures. (For a comparison of the exact free energy with the results of the Kadanoff variational and Migdal-Kadanoff methods, see [2.3,4] and [2.43], respectively). However, like other transformations in which the free energy is optimized at each renormalization step, these transformations also involve functions that are singular at the fixed point. In its present form the optimal truncation method is less useful for calculating critical exponents and derivatives of the free energy than the free energy itself.

2.5 Conclusion

Most of the work on bond-moving and variational approximations in real-space renormalization that has appeared thus far is primarily concerned with the Migdal-Kadanoff transformation, the Kadanoff lower-bound variational transformation, or the first-order cumulant transformation. The first two of these methods involve no approximations other than bond moving and thus yield lower bounds to the exact free energy. The third method is an example of an upper- bound transformation.

Among real-space renormalization procedures, the Migdal-Kadanoff method is unequalled in mathematical simplicity. It can readily be applied in arbitrary dimension to systems with continuous, as well as discrete, spins. The method works especially well just above the lower critical dimensionality, where the critical temperature is almost zero. Moreover, the method preserves the dual symmetries of certain two-dimensional Ising and Potts models and thus yields the phase diagram exactly. The corresponding critical exponents are not always very accurate, perhaps because of the inherent inconsistency in the scaling of the correlation function in the decimation transformation. From numerous applications one has a general idea about where the method can be applied with a certain amount of confidence. The Migdal-Kadanoff renormalization procedure will certainly continue to be widely used. Because of its simplicity it often constitutes the first attempt to go beyond molecular-field theory.

In the case of the first-order cumulant approximation, the upper-bound variational principle provides a criterion that eliminates some of the arbitrariness in the weight function and in the separation of the Hamiltonian into a zeroth-order part and a perturbation.

The dazzling success of the Kadanoff lower-bound transformation is predicting critical exponents when the variational parameter is fixed at the value optimizing the free energy at the fixed point is marred by the inconsistency discussed in detail in Sect.2.4. Improving the calculation of the free energy by varying the variational parameters in different applications of the transformation independently worsens the predictions for the critical exponents and leads to two fundamental dificulties from which other variational transformations also suffer: the prediction $\alpha > 0$ for the specific-heat exponent is excluded, and the renormalization tranformation involves singular functions of the coupling constants at the fixed point.

Although the reason for the success is not well understood, the single-parameter variational procedure works extremely well for a considerable number of models. It would be reassuring if some principle other than the variational principle for the free energy could be found that leads to very nearly the same value of p. A criterion that focusses attention on the critical exponents rather than the free energy would be especially welcome. An accurate free energy does not guarantee accurate critical exponents, and the optimal renormalization transformation for calculating a singular quantity is also likely to be singular.

An alternate criterion that works fairly well in predicting exponents, though not nearly so well as the Kadanoff criterion, has been suggested by KNOPS [2.85]. The work by MARTINELLI and PARISI [2.77] on improving the Migdal-Kadanoff transition, discussed at the end of Sect.2.3, suggests another possible criterion: one regards the transformation as a zeroth-order approximation and chooses p so that the first-order correction to the scaling index y_t or to some other quantity of interest vanishes. Such a criterion is, of course, considerably more difficult to apply than the original variational criterion, since the first-order corrections to the zeroth-order transformation must also be worked out.

The Migdal-Kadanoff, Kadanoff lower-bound, and first-order cumulant approximations, on which the developments discussed in this chapter are mainly based, are all more than five or six years old. Some other renormalization methods that offer definite practical advantages have appeared in the meantime. In Monte Carlo renormalization (Chap.3) the requirement of mathematical tractability, which in the case of bond-moving transformations excludes all but rather crude approximations, is considerably less stringent. In the form of finite-size scaling known as phenomenological renormalization (Chap.1), the predictions can be studied over a considerable range of finite sizes and critical exponents and couplings determined with a high precision. Bond-moving and variational approximations have played a significant role in low-dimensional critical phenomena and will certainly continue to be used. However, other effective renormalization procedures are also being devised. Whether approximations of the type considered in this chapter will constitute one of the dominant directions in future developments in real-space renormalization remains to be seen.

Acknowledgements. I have profited from useful discussions with many colleagues over a period of several years. In particular, I would like to thank Erich Eisenriegler, Hubert Knops, Dan Kroll, Marcel den Nijs, Byron Southern, and David Wallace. Reinhard Lipowsky kindly read an earlier version of this chapter and made several useful suggestions.

References

2.1 Th. Niemeijer, J.M.J. van Leeuwen: In *Phase Transitions and Critical Phenomena*, Vol.6, ed. by C. Domb, M.S. Green (Academic, New York 1976)
2.2 L.P. Kadanoff: Phys. Rev. Lett. *34*, 1005 (1975)
2.3 L.P. Kadanoff, A. Houghton, M.C. Yalabik: J. Stat. Phys. *14*, 171 (1976)
2.4 W. van Saarloos, J.M.J. van Leeuwen, A.M.M. Pruisken: Physica *92*A, 323 (1978)
2.5 M.P.M. den Nijs, H.J.F. Knops: Physica *93*A, 441 (1978)
2.6 M.N. Barber: J. Phys. A*10*, 1721 (1977)
2.7 M. Droz, S.L. Katz, J.D. Gunton: Phys. Rev. B*15*, 237 (1977)
2.8 M. Kohmoto, L.P. Kadanoff: J. Phys. A*13*, 3339 (1980)
2.9 M.C. Yalabik, A. Houghton: Phys. Lett. *61*A, 1 (1977)
2.10 T.W. Burkhardt: Phys. Rev. B*14*, 1196 (1976)
2.11 T.W. Burkhardt, H.J.F. Knops: Phys. Rev. B*15*, 1602 (1977)
2.12 C. Dasgupta: Phys. Rev. B*14*, 1221 (1976); Phys. Rev. B*15*, 3460 (1977)
2.13 T.W. Burkhardt, H.J.F. Knops, M. den Nijs: J. Phys. A*9*, L179 (1976)
2.14 M.P.M. den Nijs: Physica *95*A, 449 (1979)
2.15 S.E. Ashley, M.B. Green: J. Phys. A*9*, L165 (1976)
2.16 S.E. Ashley: J. Phys. A*11*, 2015 (1978)
2.17 B. Nienhuis, E.K. Riedel, M. Schick: J. Phys. A*13*, L31, L189 (1980); Phys. Rev. B*23*, 6055 (1981)
2.18 T.W. Burkhardt: Z. Phys. B*39*, 159 (1980)
2.19 H.J.F. Knops: Physica *93*A, 427 (1978)
2.20 M. Droz, A. Malaspinas: J. Phys. C*11*, 2729 (1978)
2.21 V.R. Vieira, J.A. Hertz: Phys. Rev. B*21*, 2336 (1980)
2.22 S.L. Katz, J.D. Gunton: Phys. Rev. B*16*, 2163 (1977)
2.23 Y.M. Shih, D.C. Jou, C.K. Pan, W.S. Lee, W.D. Chen, H.M. Huang, C.K. Hu, K.G. Chen, H.C. Tseng: Phys. Rev. B*19*, 529 (1979)
2.24 Y.M. Shih, D.C. Jou, C.K. Pan, W.D. Chen, H.M. Huang, C.K. Hu, K.G. Chen, W.S. Lee, D.S. Wan, H.C. Tseng: Phys. Rev. B*21*, 299 (1980)
2.25 S.L. Katz, M. Droz, J.D. Gunton: Phys. Rev. B*15*, 1597 (1977)
2.26 N. Jan, A.M. Glazer: Physica *91*A, 461 (1978)
2.27 B.W. Southern: J. Phys. A*11*, L1 (1978)
2.28 B. Schuh: Z. Phys. B*31*, 55 (1978)
2.29 M.J. de Oliveira, D. Furman, R.B. Griffiths: Phys. Rev. Lett. *40*, 977 (1978)
2.30 T.W. Burkhardt, E. Eisenriegler: Phys. Rev. B*17*, 318 (1978)
2.31 T.W. Burkhardt, R.H. Swendsen: Phys. Rev. B*13*, 3071 (1976)
2.32 T.W. Burkhardt, W. Kinzel: Phys. Rev. B*20*, 4730 (1979)
2.33 A.A. Migdal: Sov. Phys. JETP *42*, 743 (1976)
2.34 A.A. Migdal: Sov. Phys. JETP *42*, 413 (1976)
2.35 L.P. Kadanoff: Ann. Phys. (N.Y.) *100*, 359 (1976)
2.36 M.N. Barber: J. Phys. A*10*, 1187 (1977)
2.37 S.J. Shenker, L.P. Kadanoff, A.M.M. Pruisken: J. Phys. A*12*, 91 (1979)
2.38 L.P. Kadanoff, A. Houghton: Phys. Rev. B*11*, 377 (1975)
2.39 D.J. Wallace, R.K.P. Zia: Phys. Rev. Lett. *43*, 808 (1979)
2.40 D. Forster, A. Gabriunas: Phys. Rev. A*23*, 2627 (1981); A*24*, 598 (1981)
2.41 A.M. Polyakov: Phys. Lett. *59*B, 79 (1975)

2.42 E. Brézin, J. Zinn-Justin: Phys. Rev. Lett. *36*, 691 (1976); Phys. Rev. B*14*, 3110 (1976)
2.43 R. Lipowsky, H. Wagner: Z. Phys. *42*, 355 (1981)
2.44 B. Nienhuis, M. Nauenberg: Phys. Rev. Lett. *35*, 477 (1975)
2.45 M. Blume, V.J. Emery, R.B. Griffiths: Phys. Rev. A*4*, 1071 (1971)
2.46 V.J. Emery, R.H. Swendsen: Phys. Lett. *64*A, 325 (1977)
2.47 M. Kaufman, R.B. Griffiths, J.M. Yeomans, M.E. Fisher: Phys. Rev. B*23*, 3448 (1981)
2.48 D. Andelman, A.N. Berker: J. Phys. A*14*, L91 (1981)
2.49 G. Forgács, A. Zawadowski: Unpublished
2.50 A.N. Berker, S. Ostlund: J. Phys. C*12*, 4961 (1979)
2.51 M. Kaufman, R.B. Griffiths: Phys. Rev. B*24*, 496 (1981)
2.52 J.V. José, L.P. Kadanoff, S. Kirkpatrick, D.R. Nelson: Phys. Rev. B*16*, 1217 (1977)
2.53 J.L. Cardy, D.J. Scalapino: Phys. Rev. B*19*, 1428 (1979)
2.54 A.N. Berker, D.R. Nelson: Phys. Rev. B*19*, 2488 (1979)
2.55 V.J. Emery, R.H. Swendsen: Phys. Rev. Lett. *39*, 1414 (1977)
2.56 E. Domany, E.K. Riedel: Phys. Rev. Lett. *40*, 561 (1978); Phys. Rev. B*19*, 5817 (1979)
2.57 A.N. Berker, S. Ostlund, F.A. Putnam: Phys. Rev. B*17*, 3650 (1978)
2.58 S. Ostlund, A.N. Berker: Phys. Rev. Lett. *42*, 843 (1979)
2.59 R.M. Hornreich, R. Liebmann, H.G. Schuster, W. Selke: Z. Phys. B*35*, 91 (1979)
2.60 B.W. Southern, A.P. Young: J. Phys. C*10*, 2179 (1977)
2.61 B.W. Southern, A.P. Young, P. Pfeuty: J. Phys. C*12*, 683 (1979)
2.62 S. Kirkpatrick: Phys. Rev. B*15*, 1533 (1977)
2.63 C. Jayaprakash, J. Chalupa, M. Wortis: Phys. Rev. B*15*, 1495 (1977)
2.64 C. Jayaprakash, E.K. Riedel, M. Wortis: Phys. Rev. B*18*, 2244 (1978)
2.65 Y. Gefen, B.B. Mandelbrot, A. Aharony: Phys. Rev. Lett. *45*, 855 (1980)
2.66 M. Suzuki, K. Sogo, I. Matsuba, H. Ikeda, T. Chikama, H. Takano: Prog. Theor. Phys. *61*, 864 (1979)
2.67 M. Droz: Phys. Lett. *73*A, 407 (1979)
2.68 J.O. Indekeu, A. Stella: Phys. Lett. *78*A, 160 (1980)
2.69 M. Suzuki, H. Takano: Phys. Lett. *69*A, 426 (1979)
2.70 M. Barma, D. Kumar, R.B. Pandey: J. Phys. C*12*, L909 (1979)
2.71 C. Castellani, C. Di Castro, J. Ranninger: Nucl. Phys. B*200*, 45 (1982)
2.72 S. Caracciolo, P. Menotti: Nucl. Phys. B, to be published
2.73 M. Nauenberg, D. Toussaint: Nucl. Phys. B*190*, 217 (1981)
2.74 T.W. Burkhardt, B.W. Southern: J. Stat. Phys. *20*, 247 (1979)
2.75 T.W. Burkhardt: Phys. Rev. B*20*, 2905 (1979)
2.76 L.S. Lucena, W. Klein, R.K.P. Zia: To be published
2.77 G. Martinelli, G. Parisi: Nucl. Phys. B*180*, 201 (1981)
2.78 G. Martinelli, P. Menotti: Nucl. Phys. B, to be published
2.79 T.W. Burkhardt: Phys. Rev. B*13*, 3187 (1976)
2.80 L.P. Kadanoff: J. Stat. Phys. *15*, 263 (1976)
2.81 J.C. Le Guillou, J. Zinn-Justin: Phys. Rev. Lett. *39*, 95 (1977); Phys. Rev. B*21*, 3976 (1980)
2.82 B. Nienhuis, A.N. Berker, E.K. Riedel, M. Schick: Phys. Rev. Lett. *43*, 737 (1979)
2.83 J.L. Black, V.J. Emery: Phys. Rev. B*23*, 429 (1981)
2.84 R.B. Pearson: Phys. Rev. B*22*, 2579 (1980)
2.85 H.J.F. Knops: Physica *86*A, 448 (1977)

3. Monte Carlo Renormalization

R. H. Swendsen

With 4 Figures

A description is given of methods for investigating critical phenomena by combining Monte Carlo computer simulations of thermodynamic systems and real-space renormalization-group analyses.

3.1 Introduction

Each of the many methods developed for the study of thermodynamic phenomena and, in particular, for the calculation of critical properties has its own particular advantages and disadvantages. The real-space renormalization-group (RG) methods discussed in this volume have the important advantages of conceptual clarity and calculational simplicity, as well as having proved capable of producing excellent agreement with known results and non-trivial predictions for systems with unknown or partly known properties [3.1-4]. The main disadvantage of this approach is the uncertainty implicit in the usual truncation approximations. This feature makes it difficult to judge the accuracy and reliability of a given calculation, other than by comparison with results obtained by other methods. Although a great deal of progress has been made in the use of these methods, and confidence has increased with accumulated experience, these calculations still lack internal consistency tests for deciding when the results can be trusted [3.1,2].

By contrast, the standard Monte Carlo (MC) approach of directly simulating the model of interest produces excellent results for the properties of finite systems, with reliable internal checks [3.5-9]. For the most part, only statistical errors remain, which are well understood and whose magnitude can be calculated from the data generated by the simulation even without reference to the degree of agreement with known results. However, when an attempt is made to calculate the properties of an infinite system very close to the critical temperature, it runs into the problem of finite-size effects [3.7-25]. Since the correlation length diverges as the critical temperature is approached, it eventually becomes greater than the linear dimensions of the system being simulated. The various singularities of interest at a critical point are therefore rounded off in the finite systems simulated by MC methods, preventing us from taking much useful data extremely close to T_c.

In recent years, the combination of Monte Carlo computer simulations and the real-space renormalization-group formalism, first suggested by MA in 1976 [3.26], has been demonstrated to provide practical ways of avoiding the difficulties of both methods and producing accurate and reliable results with an efficient use of computer time.

There are two basic approaches currently in use that come under the general heading of "Monte Carlo renormalization." The simpler of the two is the "large-cell renormalization group," first introduced by FRIEDMAN and FELSTEINER [3.27] and LEWIS [3.28] and developed by REYNOLDS et al. [3.29,30], which is discussed in Sect.3.3. It is very closely related to the theory of finite-sized scaling [3.10-25] and consequently provides insight into the fundamental assumptions of theories of the critical point. The second method, which I shall refer to as MCRG [3.31-43], is described in Sect.3.4. It is more directly related to the usual real-space methods of renormalization-group analysis discussed elsewhere in this volume. The notation will be fairly standard [3.1,31] and is introduced for completeness in Sect.3.2, along with a brief summary of the necessary equations of the real-space renormalization-group formalism.

Since a number of good descriptions of standard Monte Carlo simulations exist [3.6-9] (in particular the recent review article by BINDER [3.9]), I shall not repeat the details of the method here. The most important property of MC computer simulations necessary for current purposes is that they provide an efficient method for the calculation of any correlation functions of interest that can be defined on a finite system. The only really serious limitation that prevents the standard MC techniques from providing accurate results for critical properties is the finite-size effect, which makes the extrapolation from finite to infinite systems so difficult. It is, however, worth stressing that there are several technical aspects of Monte Carlo computer simulations that are extremely important in producing reliable numbers. BINDER has provided a detailed discussion of these essential points [3.9], all of which are also important in the calculations discussed here.

Section 3.5 contains a discussion of the work that has been done to date using the MCRG method, Sect.3.6 discusses other approaches for combining Monte Carlo simulations with a renormalization-group analysis, and Sect.3.7 concludes with a few comments and speculations on the directions of future work in this field.

3.2 Basic Notation and Renormalization-Group Formalism

The methods we consider retain the advantage of being applicable to a large class of physical models, which is characteristic of both real-space RG truncation approximations and standard MC simulations. The main current requirement is that the operators be classical, although this may not be a limitation in the future [3.44-

52]. Work to date has also been restricted to cases in which the operators are associated with lattice sites. We shall denote the operator on lattice site i by σ_i. The simplest example is that of Ising spins, for which σ_i takes on only the values +1 and -1. However, the operators can take on any values without causing any particular difficulties, including an arbitrary number of discrete states (as in the q-state Potts model), a continuum of states, or even vectors, as in the planar and Heisenberg models. The Hamiltonian can be written in the general form

$$H = \sum_{\alpha} K_\alpha S_\alpha \quad , \tag{3.1}$$

where the S_α's are the various possible combinations of the σ_i's that occur in models of interest or are generated by the renormalization-group transformations. An important simple example is the operator describing the nearest-neighbor coupling:

$$S_{nn} = \sum_{<ij>} \sigma_i \sigma_j \quad . \tag{3.2}$$

The equilibrium probability distribution can then be written as

$$P(\sigma) = \exp[H(\sigma)]/Z \quad , \tag{3.3}$$

where Z is the partition function,

$$Z = \mathop{\mathrm{Tr}}_{\sigma} \exp[H(\sigma)] \quad . \tag{3.4}$$

Correlation functions are defined in the usual way by the trace over this equilibrium distribution,

$$<S_\alpha> = \mathop{\mathrm{Tr}}_{\sigma} S_\alpha(\sigma)P(\sigma) \quad . \tag{3.5}$$

The renormalization-group approach to problems of the critical behavior of thermodynamic systems attempts to transform the problem in such a way as to focus attention directly onto the critical properties of interest and thereby simplify calculations. Since the critical singularities should be least strongly influenced by the details of the behavior of the system on a short length scale, RG transformations first integrate out some fraction of the variables associated with short-wavelength fluctuations, transforming the original system into a new one with fewer degrees of freedom. Several specific transformations will be discussed, but they are generally characterized by a local grouping of the spin variables on neighboring sites into "blocks," and assigning a value to each "block spin" on the basis of the values taken on by the spins in each block. The block-spin transformations reduce the linear dimensions of the system by a factor b, where b^d is equal to the number of spins in a block. A typical example for the d = 2 Ising model would be to divide a square lattice into 2×2 blocks (b = 2) and assign the block spin σ_i' the value of +1 if the sum of the spins is positive, and -1 if it is negative. The configurations for which the sum is zero can be handled in several ways, such as assigning them to +1 and -1 with equal probability. This particular "majority-rule" transformation will be used as an example of detailed calculations in Sect.3.4.

Formally, the RG transformation can be written in terms of the equilibrium probabilities as

$$P'(\sigma') = \operatorname*{Tr}_{\sigma}\{T(\sigma',\sigma)P(\sigma)\} \quad . \tag{3.6}$$

The new probability distribution can then be interpreted in terms of an effective Hamiltonian for the renormalized block spins

$$P'(\sigma') = \exp[H'(\sigma')]/Z' \quad , \tag{3.7}$$

and this renormalized Hamiltonian can be parameterized in terms of a new set of coupling constants $\{K'_\alpha\}$.

A problem arises at this point. Even for very simple models with only nearest-neighbor interactions and simple RG transformations, such as the majority-rule transformation mentioned above, the renormalized Hamiltonian is extremely complicated in the sense that an infinite number of coupling constants are necessary to describe it completely. The effect of the renormalization-group transformation therefore involves a mapping of an infinite-dimensional space of coupling constants onto itself. The calculational difficulties this creates are important in all discussions of real-space renormalization-group methods and are solved in very different ways in different formulations of Monte Carlo renormalization- group calculations.

Since the block-spin transformation is designed to integrate out short-wavelength fluctuations while distorting the long-wavelength correlations as little as possible, it is reasonable to assume that the correlation length ξ in units of the original lattice constant a is unchanged by the transformation. This implies that in units of the new lattice constant a' the correlation length is reduced by the scale factor b,

$$\xi' = \xi/b \quad . \tag{3.8}$$

If we start with any Hamiltonian that is at criticallity in the sense of having an infinite correlation length, the renormalized Hamiltonian will also have an infinite correlation length and lie on the hypersurface of critical points in an infinite-dimensional space of coupling constants. Further iterations of the RG transformation will continue to produce Hamiltonians on the critical hypersurface and, if the transformation is well behaved, the sequence of Hamiltonians will converge towards a fixed point.

Figure 3.1 shows a highly schematic diagram of this process, which we shall use later to discuss some features of MCRG calculations. The vertical axis represents the most important coupling in the original problem. The horizontal axis serves to represent all other parameters that describe the renormalized Hamiltonians. For simplicity, I have assumed that the initial problem involved only nearest-neighbor coupling and that $H^{(0)}$ lies on the vertical axis at the intersection

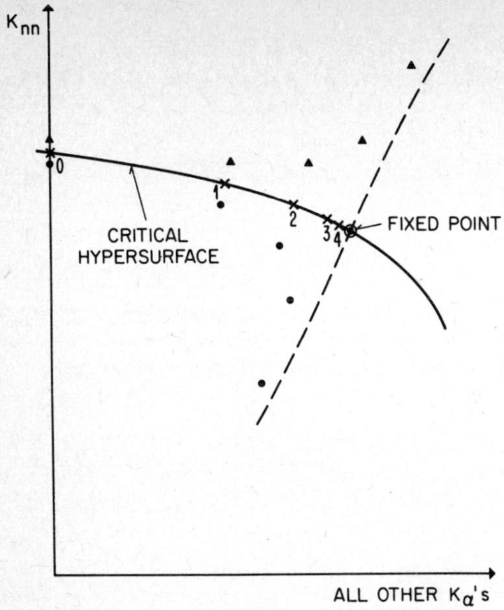

Fig.3.1. Schematic diagram of the trajectories in a many-dimensional space of coupling constants under a typical RG transformation [3.36]. The vertical axis represents the nearest-neighbor coupling constant, and the horizontal axis represent all other coupling constants generated by the RG transformation

with the critical hypersurface. The RG transformation then moves the Hamiltonian successively to the crosses labeled 1, 2, 3, etc. towards the fixed point.

We also assume that the renormalized coupling constants are analytic functions of the original coupling constants, so that if we start very slightly above or below the critical hypersurface (as indicated by the triangle and dot on the vertical axis in Fig.2.1), the renormalized Hamiltonian will initially stay close to the critical hypersurface and move towards the fixed point. Eventually, however, repetition of the RG transformation reduces the renormalized correlation length, and results in a trajectory that moves away from the critical hypersurface and away from the fixed point along a direction indicated by the dashed line. Typical sequences of renormalized Hamiltonians are indicated in Fig.3.1 by the triangles and dots. The values of the critical exponents are determined by how rapidly each RG transformation brings the renormalized Hamiltonian away from the fixed point.

If we had started very close to criticallity, the RG trajectory would have passed very close to the fixed point. We can then linearize the renormalization-group transformation about the fixed point to calculate the asymptotic behavior and the critical exponents,

$$K_\alpha^{(n+1)} - K_\alpha^* = \sum_\beta T_{\alpha\beta}\left(K_\beta^{(n)} - K_\beta^*\right) \quad , \tag{3.9}$$

where

$$T_{\alpha\beta} = \partial K_\alpha^{(n+1)}/\partial K_\beta^{(n)} \tag{3.10}$$

is evaluated at the fixed point of the RG transformation.

The critical exponents are then determined from the eigenvalue equation

$$\sum_{\beta} T_{\alpha\beta}\phi_{\beta}^{r} = \lambda\phi_{\alpha}^{r} \quad . \tag{3.11}$$

The largest eigenvalue will typically have a value greater than one and is called "relevant." It describes how rapidly the renormalized Hamiltonians move away from the fixed point. The eigenvalue exponent y_T, defined by

$$\lambda = b^{y_T} \tag{3.12}$$

gives the critical exponent $\nu = 1/y_T$, which describes the divergence of the correlation length as a function of temperature,

$$\xi = |K - K_c|^{-\nu} \quad . \tag{3.13}$$

If the initial Hamiltonian deviated from criticality because of the presence of a symmetry-breaking magnetic field, this would give rise to a different eigenvalue exponent, y_H, which would give the value of the critical exponents η and δ. All other critical exponents can then be found from the scaling equations in the usual way [3.1].

This brings us to the question of how to do practical calculations with this formalism, and the problem mentioned earlier of the infinite number of couplings generated by the block-spin RG transformations. The usual methods of real-space renormalization-group analysis attack this problem by making approximations that eliminate all but a finite number of interactions. The justification of these "truncation" approximations is generally based on the assumption that the renormalized coupling constants are short-ranged; even though all interactions are expected to have nonzero strength, most of them will be so small that their effect can be neglected. On the whole, this assumption has received considerable support from successful numerical calculations. But along with approximations that have proven to be very accurate for some systems, there are equally plausible approximations that give very poor results. Or a transformation can be good for one system and poor for another, when compared with known results. It is this uncertainty that presents problems when truncation approximations are applied to systems with unknown properties.

In principle, the usual real-space renormalization-group approximations can be systematically improved by including more and more coupling constants in the calculation. In practice this is generally impossible, because of the rapid increase in the difficulty of computation when more couplings are included [3.1].

Probably the most important advantage of combining Monte Carlo simulations with the renormalization-group analysis is that it allows all approximations to be improved systematically without a major increase in effort.

3.3 Large-Cell Monte Carlo Renormalization Group

The first effective way of applying the combination of Monte Carlo and renormalization-group methods that was developed after MA's initial suggestion [3.26] was the large-cell renormalization transformation introduced by FRIEDMAN and FELSTEINER [3.27] and LEWIS [3.28]. The method has been used widely in the study of percolation phenomena following the work of REYNOLDS et al. [3.29,30]. In addition to its usefulness, the method is particularly interesting because it demonstrates a very close connection between renormalization-group concepts and finite-size scaling.

The original presentation of the method was in the language of the renormalization group. Consider for simplicity a two-dimensional Ising model with nearest-neighbor interactions described by the Hamiltonian

$$H = K \sum_{<ij>} \sigma_i \sigma_j = KS_{nn} \tag{3.14}$$

for a finite system with dimensions $L \times 2L$. A "renormalization-group" transformation using "large" cells of dimension $L \times L$ is now defined in which the value of the block spin for each cell is given by the majority rule; each block spin takes on the value +1 or -1, depending on whether the magnetization in the corresponding cell is positive or negative. We can now consider the effective coupling between the block spins σ_1' and σ_2' defined by the simple Hamiltonian

$$H = K' \sigma_1' \sigma_2' \quad . \tag{3.15}$$

The magnitude of the effective coupling constant K' can be calculated directly from a Monte Carlo simulation of the original system through the equation

$$<\sigma_1' \sigma_2'> = \tanh(qK') \tag{3.16}$$

with $q = 1$ or $q = 2d$ depending on how we wish to interpret the transformation. (Note that all couplings other than this "nearest-neighbor" interaction are excluded since we only have two spins left in the problem, bypassing the problem of generating an infinite number of coupling constants.) At this point, the weakest step in the argument appears, and we take this "renormalized" coupling to define a Hamiltonian of the form in (3.14) for a new system of the same dimensions ($L \times 2L$) as the original system. Postponing discussion of this step, we can now use the renormalization-group formalism, look for a fixed point of the transformation from K to K', and use the derivative

$$\frac{dK'}{dK} = L^{y_T} \tag{3.17}$$

to find the thermal eigenvalue exponent y_T (the scale factor $b = L$ for this method). The derivatives can be found either by simulating a number of systems with different values of K and numerically differentiating the results or by calculating dK'/dK from the fluctuations

$$\frac{dK'}{dK} = \langle\sigma_1'\sigma_2' S_{nn}\rangle - \langle\sigma_1'\sigma_2'\rangle\langle S_{nn}\rangle \quad , \tag{3.18}$$

where we have taken a formal derivative with respect to the nearest-neighbor coupling as an example.

The critical exponents calculated in this way are then argued to converge to the true values in the limit $L \to \infty$ when the fluctuations become critical, and the details of what happens at the boundaries between the blocks becomes unimportant. This procedure does, in fact, work quite nicely, as seen from the numerical results [3.27-30,40,53-59], although the convergence to the true values of the critical exponents is quite slow and was observed to go as the $1/\ln L$ [3.27-30].

The essential equations needed for this method can be rederived from the assumptions of finite-sized scaling, without reference to the renormalization group [3.29,30]. Consider a correlation function $f(K,L)$ defined on a finite system with, for simplicity, a single coupling constant K and a characteristic linear dimension L. We restrict our attention to functions that have a finite discontinuity at the critical coupling K_c but are analytic for finite L. In particular, the large-cell transformation just discussed obviously provides such a function with $f(K, \infty)$ being zero above the critical temperature and unity below. If we make a finite-size scaling assumption, we can define a new function

$$f(K,L) \to \hat{f}((K - K_c)L^{1/\nu}) \quad , \tag{3.19}$$

where the usual power of L multiplying the scaling function is absent because $f(K,L)$ is discontinuous at the transition in the limit of $L \to \infty$. Now, consider the intersection of $f(K,L)$ with a straight line (or with a function that is analytic at K_c and can be treated as a straight line close to K_c),

$$\hat{f}(K,L) = a(K - K_c) + b \quad . \tag{3.20}$$

Define

$$x = (K - K_c)L^{1/\nu} \tag{3.21}$$

and expand f about the value x_0 for which

$$\hat{f}(x_0) = b \quad . \tag{3.22}$$

This gives us

$$\frac{d\hat{f}(x)}{dx}(x - x_0) = a(K - K_c) \quad , \tag{3.23}$$

which leads to

$$K_L = K_c + x_0 L^{-1/\nu} \tag{3.24}$$

for the intersection for a lattice of characteristic linear dimension L. If we identify the "renormalized" coupling constant K' with $f(K,L)$ and follow (3.17) to define an "effective" eigenvalue exponent y_{eff} by

$$y_{\mathrm{eff}} = \ln\left(\frac{df(K,L)}{dK}\right)/\ln L \quad , \tag{3.25}$$

we find

$$y_{\mathrm{eff}} = \nu^{-1} + \ln \hat{f}'(x_0)/\ln L \quad . \tag{3.26}$$

Both (3.24) and (3.26) agree with those derived in [3.27-30] as the basis of the large-cell renormalization-group calculations. Although the convergence indicated by these equations is not particularly rapid, the fact that its form is known for the case of simple critical points allows the use of (3.24,26) as part of the method to extrapolate from the properties of finite systems to the desired properties of the corresponding infinite systems. Numerical calculations have demonstrated that the extrapolated values do converge well in many cases. The particularly simple connection between this application of the renormalization-group approach and the older ideas of thermodynamic scaling is helpful in understanding the relationships of the concepts involved.

The extension of the large-cell method to problems of percolation by REYNOLDS et al. [3.29,30] replaces the coupling constant K by a bond or site probability p, using the probability that a finite system is spanned by a single cluster for the function f(K,L) defined above. Application of the large-cell renormalization-group method to percolation [3.29,30,53-55] and polymer [3.56-59] problems has been particularly fruitful and work in these areas is continuing.

The major limitation of the large-cell method is the small amount of information extracted from each configuration generated by the Monte Carlo simulation. To determine the critical exponent ν for an Ising model, each configuration provides only a single bit of information: the block spins are either parallel (+1) or antiparallel (-1). The same is true for a typical problem in percolation. Other information that may be contained in the MC configurations is discarded, which is unfortunate considering the large amounts of computer time necessary to generate them.

3.4 MCRG

3.4.1 Calculation of Critical Exponents

The other method of combining Monte Carlo simulations with a renormalization-group analysis is quite similar in concept to the real-space methods discussed elsewhere in this volume. Instead of using transformations that are only valid in the limit of infinite cell size, the MCRG method follows the standard real-space renormalization-group practice of using small cells throughout the calculation. This means that the many coupling constants generated by the renormalization transformation must be dealt with directly, instead of being circumvented by reducing the entire

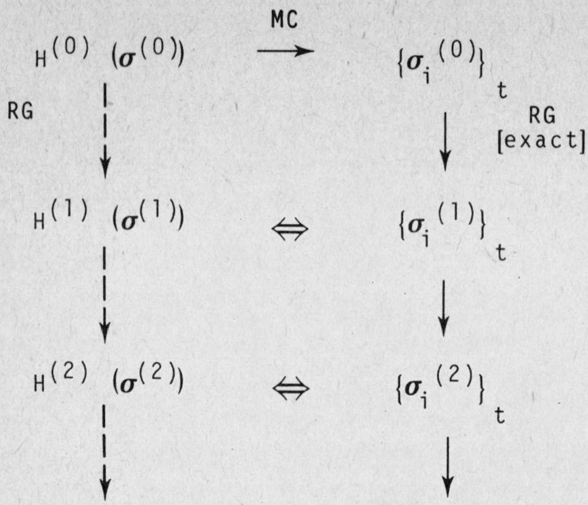

Fig.3.2. Diagram of the MCRG approach to the calculation of critical properties [3.36]

system to only two block spins. In the MCRG method, even after the renormalization transformation, it is important that there are still many spins and many effective interactions left in the system.

The strategy used in MCRG calculations is illustrated in Fig.3.2. Starting from the Hamiltonian under consideration, the usual procedure is to try to calculate the effect of an RG transformation by computing the renormalized coupling constants as a function of the original couplings, as indicated by the vertical arrows leading down from $H^{(0)}$ to $H^{(1)}$, etc. This runs immediately into the difficulty of dealing with a large number of coupling constants. Instead, we use an MC simulation to generate a sequence of configurations characteristic of the original Hamiltonian on a finite lattice. Any correlation functions of interest can then be calculated from the configurations. Since the explicit configurations are stored in the computer, it is a simple matter to use any reasonable RG transformation to divide the lattice into blocks and apply the exact RG transformation directly to these configurations. The result of applying the RG transformation is then a sequence of configurations for the renormalized spins that is characteristic of the renormalized Hamiltonian. There is no approximation involved in the RG transformation other than the statistical errors implicit in using a finite number of MC steps. The effect of all coupling constants that will fit on the lattice is automatically taken into account correctly.

The next step is to use the effective renormalized Hamiltonian on the finite lattice as an approximation to the full renormalized Hamiltonian on an infinite lattice. This is the first place in which a truncation is made in the MCRG method. Only interactions that fit on the finite lattice are included in the approximation for the infinite lattice. The number of interactions included is much larger than in any of the usual truncation approximations, since many thousands of interactions fit on a finite system of very modest dimensions. For the approximation to be valid, the effective range of the renormalized Hamiltonians, especially in the neighborhood of the fixed point, must be small compared with the dimensions of the lattice. Whether or not this is actually the case can be determined rather easily on the basis of the MCRG data by repeating the calculation for different size lattices, as demonstrated below. This provides a simple, self-consistent test of the validity of the truncation due to the finite lattice.

In MA's original paper, he attempted a direct simulation of the fixed-point Hamiltonian. However, it has proven considerably more efficient in most cases to perform an MC simulation of the system of interest at its critical temperature instead. This generally involves a simpler MC algorithm, because fewer coupling constants are needed, and does not introduce an unnecessary additional truncation, which would be implicit in parameterizing an approximate fixed point. By performing the MC simulation on a simple model at criticality, we take advantage of the property of the RG transformation discussed in Sect.3.2: repeated iterations move the renormalized Hamiltonians towards the fixed point. This gives us a sequence of approximations for the critical exponents, which converges towards the fixed point values as we iterate the RG transformation.

The matrix $T_{\alpha\beta}$ is found numerically by solving the set of chain-rule equations

$$\frac{\partial\langle S_\gamma^{(n+1)}\rangle}{\partial K_\beta^{(n)}} = \sum_\alpha \frac{\partial K_\alpha^{(n+1)}}{\partial K_\beta^{(n)}} \frac{\partial\langle S_\gamma^{(n+1)}\rangle}{\partial K_\alpha^{(n+1)}} \quad . \tag{3.27}$$

The derivatives in (4.1) are obtained from MC correlation functions,

$$\frac{\partial\langle S_\gamma^{(n+1)}\rangle}{\partial K_\beta^{(n)}} = \langle S_\gamma^{(n+1)} S_\beta^{(n)}\rangle - \langle S_\gamma^{(n+1)}\rangle\langle S_\beta^{(n)}\rangle \tag{3.28}$$

and

$$\frac{\partial\langle S_\gamma^{(n)}\rangle}{\partial K_\alpha^{(n)}} = \langle S_\gamma^{(n)} S_\alpha^{(n)}\rangle - \langle S_\gamma^{(n)}\rangle\langle S_\alpha^{(n)}\rangle \quad . \tag{3.29}$$

Note that the expectation values of certain operators, such as the magnetization in our Ising model example, vanish by symmetry on a finite lattice. This information can be used to improve the accuracy of the calculation by eliminating the second terms in (3.28,29).

The set of linear equations (3.27) is then solved for the matrix $T_{\alpha\beta}$. The eigenvalues of this matrix then give estimates of the critical exponents if the effective Hamiltonians are close to the fixed point, as discussed in Sect.3.2.

In principle, the matrix $T_{\alpha\beta}$ has an infinite number of components. Since we can only calculate a small part of this matrix, a second truncation is made at this point. However, MCRG has the important advantage that the effect of including more interactions in the calculation of the matrix can be systematically investigated. Practical calculations can include 20 interactions in the analysis without any numerical difficulties. The analysis is then usually repeated several times to see how estimates of the critical exponents are affected by the inclusion of additional interactions.

To illustrate how MCRG calculations work in practice for a simple model, I shall discuss a systematic study of the d = 2 Ising model, using a block-spin transformation with scale factor b = 2. The renormalized block-spin value was determined by majority rule, with ties being assigned values of +1 and -1 with equal probability, as determined by a random number generator. The interactions used in the RG analysis aregiven in Table 3.1. Lattice sizes ranging from 4×4 to 64×64 were used with MC simulation times given in Table 3.2.

Table 3.1. Coupling constants used for the MCRG analysis of the d = 2 Ising model

Even couplings:

Nc	Description
1	Nearest neighbor (10)
2	Next-nearest neighbor (11)
3	Four-spin coupling around a plaquette
4	Third neighbor (20)
5	Fourth neighbor (21)
6	Four-spin coupling on a sublattice plaquette
7	Fifth neighbor (22)

Odd couplings:

Nc	Description	
1	Magnetic field	
2	Three spins on a plaquette	(00)-(10)-(11)
3	Three spins in a row	(00)-(10)-(20)
4	Three spins at an angle	(00)-(10)-(21)

Table 3.2. MC simulation data for MCRG calculations. d = 2 Ising model. Data taken for correlation functions every 10 MC-steps/site. The units are 10^4 MC-steps/site

Lattice size (L)	64	32	16	8	4
Length of MC simulation	100	160	260	220	220
Steps discarded to reach equilibrium	10	6	9	2	2

Table 3.3 shows the MCRG results for the thermal eigenvalue exponent y_T for all lattices used. Each column presents data from a different MC simulation. The data are grouped according to the number (Nr) of RG iterations performed. Nc gives the number of interactions taken into account in the RG analysis, which is also the rank of the part of the matrix $T_{\alpha\beta}$ used for each numerical calculation. I have intentionally chosen lattice sizes that differ by the same scale factor as used for

Table 3.3. Critical eigenvalue exponent y_T (exact value is 1.000) for the d = 2 Ising model as a function of the number of RG iterations (Nr), the number of coupling constants in the RG analysis (Nc), and the linear dimension of the lattice (L). 2 × 2 RG block transformation

Nr	Nc	Lattice size (L) 64	32	16	8	4
1	1	0.912(2)	0.904(1)	0.897(3)	0.887(3)	0.877(3)
	2	0.967(3)	0.966(2)	0.964(3)	0.965(3)	0.988(2)
	3	0.968(3)	0.968(2)	0.966(3)	0.966(3)	0.992(2)
	4	0.969(4)	0.968(2)	0.966(3)	0.969(3)	
	5	0.969(4)	0.968(3)	0.964(4)	0.964(5)	
	6	0.969(3)	0.968(3)	0.965(4)	0.964(5)	
	7	0.969(5)	0.967(3)	0.966(4)	0.962(5)	
2	1	0.963(4)	0.953(2)	0.937(3)	0.927(3)	
	2	0.999(4)	0.998(2)	0.993(3)	1.018(3)	
	3	1.001(4)	1.000(2)	0.994(3)	1.023(3)	
	4	1.002(5)	0.998(2)	0.984(4)		
	5	1.001(5)	0.997(2)	0.980(5)		
	6	1.001(5)	0.997(2)	0.980(5)		
	7	1.000(5)	0.997(3)	0.980(5)		
3	1	0.957(2)	0.936(3)	0.921(5)		
	2	0.998(2)	0.991(3)	1.013(4)		
	3	0.999(2)	0.993(3)	1.020(3)		
	4	0.999(2)	0.987(4)			
	5	0.997(2)	0.981(7)			
	6	0.997(2)	0.979(7)			
	7	0.997(2)	0.977(9)			
4	1	0.940(7)	0.939(5)			
	2	0.993(6)	1.025(4)			
	3	0.992(6)	1.031(4)			
	4	0.988(5)				
	5	0.990(5)				
	6	0.988(5)				
	7	0.984(4)				

the RG transformation. Since the linear dimensions of a lattice are reduced by the scale factor b = 2 with each RG iteration, the 64 × 64 lattice after n iterations is reduced to the same size as the 32 × 32 lattice after n - 1 iterations. Consequently, entries along diagonal rows in Table 3.3 all refer to the same size renormalized lattices.

The numbers in parentheses give the approximate statistical uncertainty in the last digit as estimated from the spread of values obtained from MCRG analyses on groups of the MC data evaluated separately. This procedure is basically the same as the standard error analysis for MC simulations described by BINDER [3.9].

Consider the first column in Table 4.3, which gives the MCRG estimates for y_T from the 64 × 64 lattice. For the first RG step, there is a clear difference between using only a single coupling constant (nearest neighbor) in the analysis and including second-neighbor interactions. The inclusion of a four-spin coupling and more distant-neighbor interactions does not have a significant effect. It is typica of MCRG results that only a few couplings are important for the analysis, although this transformation is especially favorable in this respect. It is quite normal for three or four couplings to be sufficient for the calculation of the critical exponents and for further interactions to leave the estimates unchanged. However, there are exceptions, and this property must be established explicitly for every model investigated and every RG transformation used. For the first RG step, the estimate for y_T is about 3% below the exact value of one. This is a smaller discrepancy than found in most cases, since we cannot generally expect the original model to be so close to the corresponding fixed point. By the second RG step, the renormalized Hamiltonian has moved closer to the fixed point and the estimates of y_T have come closer to the exact value. In this particular case, deviation from the exact value is so small that it is within the statistical error of about 0.5%. The third iteration comes even closer to the fixed point and no significant change in the estimates for y_T can be seen, since it is within the statistical errors. There is a small decrease when more distant neighbors are included, but since it is only 0.3%, it is hard to know whether it is real. On the other hand, it is quite likely that some such effect is present, since the renormalization has reduced the lattice dimensions to 8×8 and interactions that extend over two lattice spacings are no longer truly short ranged with respect to the dimensions of the system. The fourth RG iteration brings the lattice down to 4×4 and it is not at all surprising that a finite-size effect is clearly visible. What is surprising is that the effect is so small (only 2%), even though the interactions used in the RG analysis now reach halfway around the system.

Comparison with the data for the 32×32 system shows that the first two RG iterations are essentially identical, with the possible exception of the same finite-size effect in the second RG iteration for the 32×32 lattice that may be present in the third iteration of the 64×64 lattice. The third iteration of the 32×32 lattice clearly shows the same finite-size effect as the fourth iteration of the 64×64 system, which is exactly as expected, since both systems have been reduced to 8×8 lattices. For the fourth iteration, the 32×32 system is reduced to a 2×2 lattice and only the first three operators are linearly independent. Even so, deviation of the estimates from the exact value is still only about 3%, which is remarkably small. Note that the size effect is not monotonic and the estimate for y_T increases for the 2×2 lattice.

The approach to the fixed point and the finite-size effect can be seen in Table 3.3 also for the smaller lattices. It is perhaps especially interesting to point out that the 4×4 lattice produces a very good estimate for y_T but that this comes from a compensation of the finite-size effect by the initial deviation from the fixed point.

Table 3.4. Critical eigenvalue exponent y_H (exact value is 1.875) and for the d = 2 Ising model as a function of the number of RG iterations (Nr), the number of coupling constants in the RG analysis (Nc), and the linear dimension of the lattice (L). 2 × 2 RG block transformation

Lattice size (L)						
		64	32	16	8	4
Nr	Nc					
1	1	1.8810(1)	1.8807(1)	1.8797(2)	1.8769(4)	1.8698(5)
	2	1.8804(1)	1.8803(1)	1.8800(2)	1.8795(4)	1.8758(5)
	3	1.8806(1)	1.8806(1)	1.8802(2)	1.8794(4)	
	4	1.8808(1)	1.8808(1)	1.8806(2)	1.8799(4)	
2	1	1.8757(2)	1.8748(2)	1.8719(2)	1.8667(4)	
	2	1.8758(2)	1.8757(2)	1.8747(2)	1.8731(5)	
	3	1.8758(2)	1.8757(2)	1.8745(2)		
	4	1.8759(2)	1.8758(2)	1.8749(2)		
3	1	1.8731(4)	1.8710(5)	1.8658(5)		
	2	1.8740(4)	1.8742(5)	1.8728(7)		
	3	1.8740(4)	1.8739(5)			
	4	1.8741(4)	1.8743(5)			
4	1	1.8706(5)	1.8662(2)			
	2	1.8735(7)	1.8723(2)			
	3	1.8732(8)				
	4	1.8737(9)				

Table 3.4 shows similar data for the magnetic eigenvalue exponent, y_H. All trends seen in Table 3.4 are also present here, although the deviations from the exact value are uniformly smaller. This effect is also characteristic of the magnetic eigenvalue, which is generally much less sensitive to size effects and the distance from the fixed point than the thermal eigenvalues. The statistical errors are also much smaller, which is partially due to the vanishing of the correlation functions in the subtracted terms in (3.28,29), so that only the first correlation functions are needed.

In principle, the eigenvalues of the matrix $T_{\alpha\beta}$ not only provide estimates for the relevant operators, but for the irrelevant ones as well. For the d = 2 Ising model, the first irrelevant eigenvalue exponent is $y_{T2} = -1$. Table 3.5 presents MCRG data for this eigenvalue, which is seen to be consistent with the exact value but with much larger error bars than were found for y_T. It is also seen that at least five interactions must be taken into account before the estimates for y_{T2} approach the value expected. Because of this, no meaningful results can be obtained for a 2 × 2 renormalized lattice and the data from a 4 × 4 lattice shows a strong finite-size effect.

Examination of the magnetic eigenvalues also shows evidence for an irrelevant eigenvalue, but its apparent value of about -0.3 is lower than the expected $y_{H2} = -0.125$ and moves slowly in the direction of the expected value as additional

Table 3.5. Critical eigenvalue exponent y_{T2} (exact value is -1.000) for the d = 2 Ising model as a function of the number of RG iterations (Nr), the number of coupling constants in the RG analysis (Nc), and the linear dimension of the lattice (L). 2 × 2 RG block transformation

		Lattice size (L)			
		64	32	16	8
Nr	Nc				
1	2	-2.19(3)	-2.13(5)	-2.14(3)	-2.10(5)
	3	-2.17(3)	-2.11(6)	-2.08(2)	-2.01(5)
	4	-1.11(3)	-1.09(2)	-1.11(4)	-1.53(7)
	5	-0.94(3)	-0.88(2)	-0.87(5)	-0.88(7)
	6	-1.04(3)	-0.98(3)	-1.01(6)	-0.92(7)
	7	-1.08(11)	-1.01(6)	-1.05(8)	-0.94(10)
2	2	-2.20(4)	-2.21(6)	-2.16(5)	
	3	-2.34(6)	-2.34(8)	-2.24(6)	
	4	-1.14(3)	-1.11(6)	-1.23(4)	
	5	-0.94(3)	-0.89(6)	-0.74(4)	
	6	-1.04(3)	-0.96(9)	-0.77(4)	
	7	-1.03(4)	-1.01(11)	-0.77(4)	
3	2	-2.20(3)	-2.16(7)		
	3	-2.56(6)	-2.26(10)		
	4	-1.16(3)	-1.41(8)		
	5	-0.91(2)	-0.76(8)		
	6	-0.98(2)	-0.78(8)		
	7	-0.99(10)	-0.73(9)		
4	2	-2.32(8)			
	3	-2.56(12)			
	4	-1.35(10)			
	5	-0.80(8)			
	6	-0.84(9)			
	7	-0.76(10)			

interactions with more factors of the spin variables are included in the analysis, indicating that the analysis is not yet sufficient.

A further test of the consistency of the MCRG results, using the same data, can be made by regarding two iterations of the 2 × 2 RG transformation as a single transformation with scale factor 4. The linearized transformation is described by the square of the $T_{\alpha\beta}$ matrix for the 2 × 2 transformation. This feature is important as a test of the effect of including many interactions in the RG analysis, because the multiplication is performed implicitly by the MCRG method, effectively including all interactions that will fit on the lattice. If there is a discrepancy between the analysis with a single and a double iteration, then the effect of the other interactions cannot be neglected. Unfortunately, it is not necessarily true that the lack of a discrepancy proves that all is well with the restriction to a small number of interactions, because an eigenvalue of interest may lie outside or partially outside the space spanned by the interactions used in the analysis.

This trick can be generalized to m RG iterations forming a single transformation with scale factor 2^m. The results of such calculations are shown in Table 3.6 for

Table 3.6. Critical eigenvalue exponent y_T (exact value is 1.000) for the d = 2 Ising model as a function of the number of RG iterations (Nr), the number of coupling constants in the RG analysis (Nc), and the power of the scale factor m. $(2\times 2)^m$ RG block transformation on a 64×64 lattice

m (scale factor: 2^m)		1	2	3	4
Nr	Nc				
1	1	0.912(2)	0.948(2)	0.961(2)	0.965(3)
	2	0.967(3)	0.983(3)	0.990(2)	0.990(3)
	3	0.968(3)	0.985(3)	0.991(2)	0.989(3)
	4	0.969(4)	0.985(3)	0.991(2)	0.989(4)
	5	0.969(4)	0.986(3)	0.989(3)	0.990(5)
	6	0.969(3)	0.987(3)	0.989(3)	0.990(5)
	7	0.969(5)	0.986(3)	0.989(3)	0.991(5)
2	1	0.963(4)	0.972(3)	0.975(3)	
	2	0.999(4)	0.999(2)	0.998(3)	
	3	1.001(4)	1.000(2)	0.997(3)	
	4	1.002(5)	1.001(3)	0.998(4)	
	5	1.001(5)	1.000(3)	0.999(5)	
	6	1.001(5)	0.999(3)	0.999(5)	
	7	1.000(5)	1.000(3)	1.000(4)	
3	1	0.957(2)	0.964(4)		
	2	0.998(2)	0.996(4)		
	3	0.999(2)	0.995(4)		
	4	0.999(2)	0.996(5)		
	5	0.997(2)	0.997(5)		
	6	0.997(2)	0.996(5)		
	7	0.997(2)	0.996(5)		
4	1	0.940(7)			
	2	0.993(6)			
	3	0.992(6)			
	4	0.988(5)			
	5	0.990(5)			
	6	0.988(5)			
	7	0.984(4)			

the 64×64 lattice. It is clear that the consistency tests are nicely satisfied for this example. Table 3.7 shows the corresponding data for the magnetic eigenvalue, which exhibits even better accuracy.

It is also possible to use finite-size scaling [3.8-25] within the context of an MCRG calculation. With each renormalization step, the linear dimensions of the lattice shrink by the scale factor b. If we are at a fixed point of the transformation, the susceptibility χ should scale as

$$\chi = \langle (S_H)^2 \rangle = L^{d+2-\eta} \quad . \tag{3.30}$$

Since the RG transformation converges towards the fixed point if we have started at a critical point, this equation gives us another method of finding a sequence of estimates that converges towards the value of η with no additional effort. Table 3.8 shows the results of such calculations for the d = 2 Ising model simu-

Table 3.7. Critical eigenvalue exponent y_H (exact value is 1.875) for the d = 2 Ising model as a function of the number of RG iterations (Nr), the number of coupling constants in the RG analysis (Nc), and the power of the scale factor m. $(2\times 2)^m$ RG block transformation on a 64×64 lattice

		m (scale factor: 2^m)			
		1	2	3	4
Nr	Nc				
1	1	1.8810(1)	1.8784(1)	1.8766(2)	1.8752(3)
	2	1.8804(1)	1.8781(1)	1.8768(2)	1.8760(3)
	3	1.8806(1)	1.8782(1)	1.8769(2)	1.8760(3)
	4	1.8808(1)	1.8784(1)	1.8770(2)	1.8763(3)
2	1	1.8757(2)	1.8744(2)	1.8732(3)	
	2	1.8758(2)	1.8750(2)	1.8746(4)	
	3	1.8758(2)	1.8750(2)	1.8745(4)	
	4	1.8759(2)	1.8750(2)	1.8747(4)	
3	1	1.8731(4)	1.8720(5)		
	2	1.8740(4)	1.8739(5)		
	3	1.8740(4)	1.8738(5)		
	4	1.8741(4)	1.8741(6)		
4	1	1.8706(5)			
	2	1.8735(7)			
	3	1.8732(8)			
	4	1.8737(9)			

Table 3.8. Estimates of the exponent η for the d = 2 Ising model from finite scaling on renormalized lattices. Exact value is $\eta = 0.25$. Values of η obtained by comparison of renormalized lattices (n) and (n-1) from a single MC simulation as a function of the size of the original lattice (L). 2×2 RG block transformation

	Lattice size (L)				
	64	32	16	8	4
n					
1	0.238	0.237	0.236	0.232	0.220
2	0.247	0.246	0.242	0.227	
3	0.250	0.244	0.229		
4	0.245	0.228			

lations. The convergence is quite good, although a small size effect is seen when the renormalized lattice is reduced to 4×4. The deviation from the correct value of 0.25 is about 8% when the lattice size is reduced to 2×2.

3.4.2 Calculation of Renormalized Coupling Constants

An interesting feature of the calculations just discussed is that the renormalized Hamiltonians are never explicitly calculated. This is a great advantage in the determination of the critical exponents, because many thousands of effective couplings can be taken into account implicitly without introducing the numerical errors that would be inevitable if we had to deal with them explicitly. On the other hand, there are two important reasons for wanting to be able to calculate the renormalized Hamiltonians with the MCRG method.

a) For the determination of the critical exponents, it is necessary to perform a Monte Carlo simulation of the system of interest at its critical temperature. For a number of systems in two dimensions, the critical temperature can be obtained exactly from duality relations [3.60-71], greatly simplifying the computations. However, for most systems of interest the critical temperature is unknown. It is possible to use the convergence of the estimates for the critical exponents as a criterion for finding the critical temperature [3.32,35], but it would be preferable to reserve it as a test of the consistency of the method. If the effect of the renormalization transformation on the coupling constants is known explicitly, the critical temperature can be calculated directly, saving the convergence of the eigenvalues for use as a consistency check.

b) Although the convergence of the RG transformations has been quite rapid for most systems studied so far, there are exceptions for which the many iterations would be necessary to come sufficiently close to the fixed point for an accurate calculation. For such cases it would be extremely helpful to be able to find the approximate position of a projection of the fixed point onto some convenient subspace of coupling constants, to make the calculation more efficient. This point of view is closely related to MA's original work, in which he suggested simulating the fixed point directly [3.26].

Recently, techniques have been developed to treat this problem [3.40,41], based on a method introduced by WILSON [3.72,73] for the study of lattice gauge theories [3.2-4,72,74]. WILSON has pointed out that the size effect can be circumvented for certain purposes by performing two simulations on lattices of different sizes. If the two systems differ in size by the scale factor b, one application of the renormalization-group transformation to the large system will produce a system equal in size to the smaller system, as shown schematically in Fig.3.3. We now have two equal-size systems, one of which is described by $H^{(1)}$ and the other by $H^{(0)}$.

Since the larger system can have a linear dimension of eight lattice constants, the correlation functions will show a large size effect. However, since both lattices are now the same size, this effect will be identical for both systems. WILSON then suggested adjusting the coupling constants used in the simulation of the smaller system until the correlation functions were the same as those obtained by renormalizing the larger system. If the coupling constants that produce agreement in the correlation functions are the same as those used in the original simulation of the large lattice, we have found an approximation to the fixed point.

Unfortunately, for problems in statistical mechanics, we have to be able to deal with many coupling constants if we are to have a generally applicable method. If we have to scan a space of many coupling constants to find the renormalized Hamiltonian,

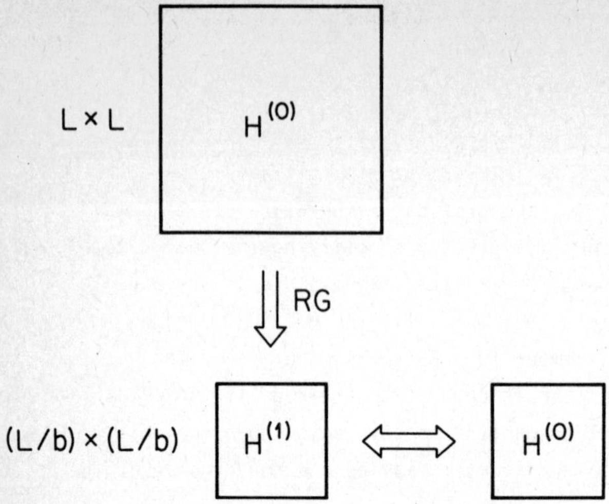

Fig.3.3. Diagram of WILSON's two-system simulation procedure [3.70], comparing the correlation functions obtained from an RG transformation of a system with original linear dimension L with a system with original linear dimension L/b

the method becomes impractical. However, we can use the differences in the correlation functions to calculate the differences in the coupling constants. If the differences in the correlation functions are not too large, we can make the linear approximation

$$\begin{aligned} \delta\langle S_\alpha^{(1)}\rangle &= \langle S_\alpha^{(1)}\rangle_L - \langle S_\alpha^{(0)}\rangle_S \\ &= \sum_\beta \frac{\partial\langle S_\alpha^{(1)}\rangle}{\partial K_\beta^{(1)}}\,\delta K_\beta \quad , \end{aligned} \tag{3.31}$$

where L and S refer to the large and small lattices. The derivatives are given by (3.29) and can be calculated from either lattice (taking the average of the two correctly includes the quadratic terms). By applying this procedure to successive iterations of an RG transformation, we can confirm that the renormalized Hamiltonians approach a fixed point.

This approach suggests a number of methods for calculating approximate fixed points. WILSON has suggested a particularly simple and efficient procedure, using the derivatives of the renormalized correlation function after n iterations, with respect to the original coupling constants [3.75].

Suppose the Hamiltonian $H^{(0)}$ is actually the fixed point of the transformation. Then $H^{(1)} = H^{(0)}$ by definition and the correlation functions must be the same on the renormalized large lattice and the small lattice. Now consider the effect of changing the coupling constants slightly to move slightly off the fixed point. Now $H^{(1)}$ is not equal to $H^{(0)}$ and the correlation functions will differ. From the nonzero differences in the correlation functions, we can calculate the deviations of the coupling constants in H from the fixed-point Hamiltonian.

The changes in the correlation functions are given by

$$\delta\langle S_\alpha^{(n)}\rangle_L = \sum_\beta \frac{\partial\langle S_\alpha^{(n)}\rangle_L}{\partial K_\beta^{(0)}}\,\delta K_\beta^{(0)} \quad , \tag{3.32}$$

where

$$\frac{\partial\langle S_\alpha^{(n)}\rangle_L}{\partial K_\beta^{(0)}} = \langle S_\alpha^{(n)}\, S_\beta^{(0)}\rangle_L - \langle S^{(n)}\rangle_L\langle S^{(0)}\rangle_L \tag{3.33}$$

with corresponding equations for the smaller lattice. By solving the set of linear equations

$$\langle S_\alpha^{(n)}\rangle_L - \langle S_\alpha^{(n-1)}\rangle_S = \sum_\beta \left[\frac{\partial\langle S_\alpha^{(n)}\rangle_L}{\partial K_\beta^{(0)}} - \frac{\partial\langle S_\alpha^{(n-1)}\rangle_S}{\partial K_\beta^{(0)}}\right]\delta K_\beta^{(0)} \quad , \tag{3.34}$$

we can determine what changes in the coupling constants are necessary to make the correlation functions match. This allows us to calculate an approximate fixed point using fairly small lattices.

Note that as the number of RG iterations n increases, the deviations of $H^{(n)}$ from the fixed point (and from $H^{(n-1)}$) in the irrelevant directions decrease. The corresponding differences in the correlation functions also decrease and become more difficult to determine accurately, so that the best determination of the fixed point is actually obtained from a single RG iteration.

On the other hand, when the original MC simulation is not performed exactly at criticality, the deviations in the relevant directions increase upon repetition of the RG transformation and the corresponding differences in the correlation functions become amplified. This feature makes the method extremely sensitive to the location of the critical temperature, since very small changes in the temperature result in large changes in the correlation functions after a few RG transformations. Since (3.34) can treat several coupling constants in a single calculation, this method can also be used to find multicritical points [3.41]. It has proven especially effective to generalize (3.34) to perform a least-squares fit of the $\{K_\alpha\}$'s to several correlation-function differences.

Table 3.9 shows the results of such a calculation of the critical temperature for the simple example of the two-dimensional Ising model. The numbers at the top of each column give the original linear dimensions of the lattices being compared. The first RG iteration shows a small deviation from zero due to the irrelevant operators. As further iterations bring the renormalized Hamiltonians closer and closer to the fixed point (Fig.3.1), the magnitude of the irrelevant contributions decreases. In this particular example, the second iteration is sufficient

Table 3.9. Estimates of the deviation of the simulated coupling constant from the true value K_c (exact value: $K_c = 0.4406879$). d = 2 Ising model. 2 × 2 RG block transformation. K_c (apparent) - K_c (true)

No	Nc	Lattice sizes being compared: 64-32	32-16	16-8	8-4
1	1	0.0024(3)	0.0029(1)	0.0037(2)	0.0036(5)
	2	0.0009(3)	0.0012(1)	0.0017(2)	0.0008(5)
	3	0.0014(3)	0.0017(1)	0.0023(2)	0.0018(6)
	4	0.0012(3)	0.0015(1)	0.0022(2)	0.0019(6)
	5	0.0005(3)	0.0006(2)	0.0012(2)	-0.0002(4)
	6	0.0005(3)	0.0006(2)	0.0012(2)	-0.0002(4)
	7	0.0005(3)	0.0006(2)	0.0012(2)	-0.0003(4)
2	1	0.0002(3)	0.0002(2)	0.0004(2)	-0.0013(2)
	2	0.0000(3)	0.0000(2)	0.0002(3)	-0.0023(3)
	3	0.0001(3)	0.0001(2)	0.0003(3)	-0.0018(3)
	4	0.0000(3)	0.0001(2)	0.0003(3)	
	5	-0.0000(3)	0.0001(3)	0.0001(3)	
	6	-0.0000(3)	0.0001(3)	0.0001(3)	
	7	-0.0000(3)	0.0001(3)	0.0001(3)	
3	1	-0.0002(4)	-0.0001(4)	-0.0000(2)	
	2	-0.0002(4)	-0.0002(4)	-0.0017(3)	
	3	-0.0002(4)	-0.0002(4)	-0.0001(3)	
	4	-0.0002(4)	-0.0002(4)		
	5	-0.0002(4)	-0.0002(5)		
	6	-0.0002(4)	-0.0002(5)		
	7	-0.0002(4)	-0.0002(5)		
4	1	-0.0003(4)	-0.0003(6)		
	2	-0.0003(5)	-0.0003(6)		
	3	-0.0003(6)	-0.0003(6)		
	4	-0.0003(6)			
	5	-0.0003(6)			
	6	-0.0003(6)			
	7	-0.0003(6)			

to reduce the systematic error due to the irrelevant operators below the statistical uncertainty, which is a few parts in [3.10]. Note that even a comparison of 8 × 8 and 4 × 4 lattices gives the correct critical coupling to better than 0.5%. Comparison of the 16 × 16 and 8 × 8 lattices shows no systematic deviation from the true critical point at the second RG iteration within the statistical errors.

3.5 MCRG Calculations for Specific Systems

The initial applications of any new method of studying phase transitions are really tests of the method and naturally involve models for which information is already available. The most obvious candidate is the two-dimensional Ising model, discussed in Sect.3.3. This is the model originally treated by MA [3.26] and SWENDSEN [3.31,35] to demonstrate the feasibility of the approach. Calculations have been performed using a rather wide variety of block-spin transformations [3.35]. Majority-rule transformations with and without tie-breakers (for the case in which the sum of the spins in a block is zero) have been used for 2×2 and 3×3 blocks. The five-spin block suggested by VAN LEEUWEN [3.76] has also been used successfully in the original form of a central spin with its four second-nearest neighbors [3.32,35], as well as in variations using the four nearest neighbors [3.35] on a square lattice and the six second-nearest neighbors on a triangular lattice [3.39]. All of these transformations produce good agreement with the exact results. This is an essential consistency test of the method. Since the same renormalization-group assumptions form the basis of every calculation and the same approximations are involved in each case (finite lattice size, finite number of Monte Carlo steps), we cannot discard transformations that give bad results, but we must insist on the results of all RG transformations being consistent. Fortunately, this has proven to be the case in all work to date, and even the decimation transformation gives consistent results, although the convergence is relatively slow because the fixed point is located at zero coupling [3.35,38].

Other models are available for which exact solutions exist. Of particular interest is the eight-vertex or Baxter model [3.77-81]. This model can be represented by Ising spins on a square lattice with

$$H = K_{nnn} S_{nnn} + K_4 S_4 \quad , \tag{3.35}$$

where S_{nnn} is the sum over all next-nearest-neighbor products and S_4 is the sum over all four-spin products around an elementary plaquette. The critical exponents do not take on fixed universal values but are functions of the coupling constants,

$$\begin{aligned} \nu &= \nu(K_4) \\ &= \pi/\{2\pi - 2\cos^{-1}[\tan(2K_4)]\} \quad . \end{aligned} \tag{3.36}$$

This behavior can be represented within the renormalization-group formalism by a fixed line, instead of the simpler fixed point discussed above. However, from a computational point of view this presents considerable difficulties for the usual truncation approximations, which have not been able to reproduce the true behavior.

SWENDSEN and KRINSKY [3.32] have shown that the MCRG method does not run into the usual difficulties and can easily reproduce the correct function $\nu(K_4)$ within 1% for ν ranging from 0.7 to 1.35. Their calculation used the five-spin block transformation suggested by VAN LEEUWEN [3.76]. This transformation has a scale factor $b = \sqrt{5}$ and has the advantage of not disturbing the symmetry of the sublattices. Their calculation also emphasized another advantage of this approach in that they were able to calculate several independent critical exponents with a single MC simulation. In addition to $y_T = 1/\nu$, they correctly obtained the values of $y_H = 1.875$ (which describes the magnetization) to within 0.3%, $y_{H2} = 0.875$ (which corresponds to a three-spin coupling) to within 4%, and $y_s = 0.75 + 0.25\, y_T$ (which corresponds to breaking the symmetry with a nearest-neighbor coupling) to within 0.5%. The only eigenvalue with which they had difficulty was the marginal eigenvalue $y_B = 0$, along the fixed line. This has proven to be due to their having not included sufficient four-spin operators in their analysis of the $T_{\alpha\beta}$ matrix. They had only one four-spin operator that coupled the two sublattices. By including two more such operators, they obtained y_B values with a magnitude of less than 0.05 [3.82]. It is possible, within the MCRG method, to investigate the eigenvectors and eigenvalues associated with any operators of interest. However, since all calculations are carried out on finite computers, it is not possible to check all eigenvectors. This means that each problem must be formulated carefully, making sure that all important operators are taken into account, to make best use of the computer time available.

SWENDSEN and KRINSKY [3.32] also investigated the d = 2 Ising antiferromagnet with nearest- and second-nearest-neighbor interactions. This model had been predicted by VAN LEEUWEN [3.76] to map onto the Baxter fixed line under renormalization and therefore to have critical exponents that depend on the value of the second-neighbor coupling K_2 when K_2 is large and negative. This model is more difficult than the Baxter model, because the location of T_c is unknown. SWENDSEN and KRINSKY [3.32] used the convergence of the magnetic eigenvalue as a criterion for determining T_c and were able to confirm VAN LEEUWEN's prediction by explicitly calculating the dependence of y_T on K_2. Later calculations [3.82] using the two-lattice simulations described in the previous section [3.41,72,75] have given the same critical temperatures as the original work, and the value of ν they found also agree with the finite-size scaling approach of NIGHTINGALE [3.24].

Another problem of interest in two dimensions, for which some exact information is available, is the q-state Potts model [3.83-96]. This model consists of variables at each lattice site that take on the values 1,2,...,q, and the Hamiltonian

$$H = K \sum_{<ij>} \delta_{\sigma_i,\sigma_j} \quad , \tag{3.37}$$

where the sum runs over all nearest-neighbor pairs. The special case of q = 2 is the usual Ising model. The critical temperature is known exactly from the duality relations [3.83] for all values of q, which simplifies MCRG calculations [3.31,36, 42]. The model is of particular interest for several reasons. One of the most important is due to BAXTER's proof that the phase transition is first order for $q > 4$ [3.86], although it is second order for $q \leq 4$. This behavior has proven extremely difficult to reproduce by truncation approximations and has only been accomplished by NIENHUIS et al. [3.89] through the use of a renormalization transformation that is not restricted to the original Potts variables but also introduces "vacancies" into the renormalized configurations. A schematic picture of the

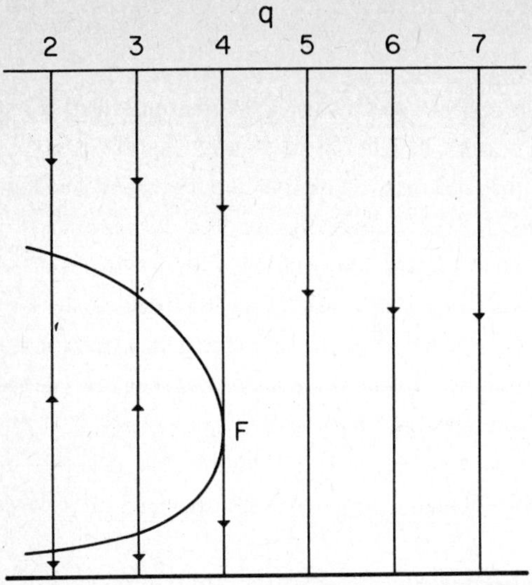

Fig.3.4. Simplified schematic diagram [3.34] of critical surface for the two-dimensional q-state Potts model, according to the suggested picture of NIENHUIS et al. [3.87]

qualitative features of their flow diagram is sketched in Fig.3.4 [3.36,89]. Assuming for simplicity that the RG transformation does not change the value of q, the RG flows move towards a fixed line at $q \leq 4$, with critical exponents that depend on the value of q [3.90,93-95]. For $q > 4$, the RG flows miss this line completely and continue on to a line of discontinuity fixed points that describe the first-order transition with eigenvalue exponent $y = d$ [3.97]. According to this picture, the value of $q = 4$ is special in that the RG flows approach the fixed line tangentially. Since there is no flow along the fixed line by definition, there must be an extra marginal operator for $q = 4$. This marginal operator has the physical consequence of predicting logarithmic corrections to the usual power law [3.36,91-96] as well as implying that the RG flows should be particularly slow. This behavior has indeed been seen [3.36,42,107]. The value of y_T is expected to be 1.5 [3.90,98], but even four iterations are not sufficient to produce convergence and the fourth estimate for y_T is only 1.34 [3.36,42]. The slow convergence does not depend on the RG transformation and the same effect is seen when vacancies are introduced into the transformation [3.36] following NIENHUIS et al. [3.89].

These results are in sharp contrast to the BAXTER-WU model [3.98], which is expected to be in the same universality class as the $q = 4$ Potts model. It consists of Ising spins on a triangular lattice, which triplet interactions on each triangle,

$$H = K \sum_{<ij>} \sigma_i \sigma_j \sigma_k \quad . \tag{3.38}$$

The model can be divided into three sublattices and is symmetric under the reversal of the spins on any two sublattices. This gives rise to four symmetric phases in the ordered state, which is the same symmetry as the $q = 4$ Potts model. However, from the exact solution of this model [3.98], we know that there is no logarithmic correction, and hence the convergence of the RG method is not expected to be particularly slow. This rapid convergence has been confirmed by NOVOTNY et al. [3.39], using a seven-spin block that is the direct generalization of VAN LEEUWEN's five-spin block transformation [3.76]. This was an important test of both the MCRG method and the predictions for the q = 4 Potts model to rule out the possibility of the slow convergence of the MCRG data being simply an artifact of the method. Even so, the most that can be said on the basis of this MCRG data is that it is consistent with the expected behavior of the q-state Potts models and not that it has been explicitly confirmed.

More support for the qualitative picture of NIENHUIS et al. [3.89] and the extended DEN NIJS conjecture for the critical exponents [3.89,90] has come from an MCRG calculation of the tricritical exponents for q = 2 by LANDAU and SWENDSEN [3.41]. The continuation of the fixed line shown in Fig. 3.4 is predicted to describe tricritical behavior [3.89] that suggests testing the DEN NIJS formula in a region far from the models on which it was originally conjectured [3.90]. The difficulty of locating the tricritical point was overcome by the method described in Sect.3.4.2 [3.38-41,72,75] and tricritical exponents were found to be in agreement with both the DEN NIJS conjecture [3.90] and the conjecture of PEARSON [3.93] and NIENHUIS et al. [3.95].

Extensions of the MCRG method to the three-dimensional simple-cubic Ising model have been made by BLÖTE and SWENDSEN [3.34] and the face-centered-cubic Ising model has been treated by SAHNI and BANAVAR [3.43]. The results of both calculations are consistent with each other and with the critical exponents obtained with other methods. In both cases, the accuracy of the calculation was about 1-2%, which is not yet competitive with older techniques. However, the outlook for the future is very good, since neither calculation used an exceptionally large amount of computer time. In addition, neither calculation took advantage of the ambiguity in defining the position of the blocks to improve the statistical independence of the renormalized configurations by redefining the blocks every time data is taken [3.36]. This trick alone can improve the accuracy of the third iteration by almost an order of magnitude. If the MC simulation could then be performed on a special-purpose computer, the number of passes through the lattice might be increased by a factor of 100, giving another order of magnitude in the statistical accuracy. The only serious difficulty seems to be the convergence towards the fixed point. The first three RG iterations gave BLÖTE and SWENDSEN [3.34] estimates for ν of 0.70, 0.64, and 0.63. If the last value is characteristic of the fixed point, all is well and we only have to improve the statistics. However, to demonstrate that this is true, we need at least one more iteration (requiring a 64×64 lattice), or a starting Hamiltonian closer to the fixed point to make three RG iterations sufficient.

BLÖTE and SWENDSEN have also performed MCRG calculations for the d = 4 Ising model [3.37], finding agreement with the predictions of mean-field theory. The accuracy of the calculation was not sufficient to detect the expected marginal operator characteristic of the crossover dimension [3.2-4].

The q = 3 Potts model has also been studied by the MCRG method in both three and four dimensions and the phase transition determined to be first order in both cases [3.33], although some questions remain as to the nature of the RG description of this transition.

3.6 Other Approaches to the Monte Carlo Renormalization Group

The flexibility of the Monte Carlo method of computer simulations and the real-space renormalization-group formalism is such that there are more possibilities than the two methods discussed above. There are at least three other approaches that have been taken, each of which has certain advantages to recommend it.

The early efforts to use this combination to develop an efficient tool for studying phase transitions considered the problem within the framework of the various perturbation expansions. The Monte Carlo simulation can simplify the calculational effort involved in evaluating the factors entering, for example, the cumulant expansion of the linearized RG equations [3.99-102], rather than eliminating the inherent approximations. Although these calculations have not produced the same accuracy or reliability as the large-cell renormalization group or the MCRG, they have contributed to the understanding of the truncation approximations. The approach is also of considerable potential importance for calculations of global properties. It may be possible to use information obtained from calculations such as those of RACZ and RUJAN [3.99], KINZEL [3.100], MÜLLER-KRUMBHAAR [3.101], and MUTO et al. [3.102] along with appropriate MCRG data to design renormalization-group truncation approximations with accuracy and reliability. The accuracy of the approximation could be checked at various points with MCRG, while the calculational simplicity of the truncation approximations would enable global phase diagrams to be computed rapidly.

The second approach that has been taken is a rather direct application of WILSON's [3.72] procedure of matching renormalized correlation functions from two MC simulations to follow the RG trajectories in a subspace of coupling constants. This has been applied successfully to the numerically difficult problem of the two-dimensional Heisenberg ferromagnet by SHENKER and TOBOCHNIK [3.73], as well as to the four-dimensional SU(2) and U(1) lattice gauge models by CREUTZ [3.74]. This is an especially useful approach for systems not expected to have a phase transition, where the RG trajectories can be shown to flow in only one principal direction for all initial coupling constants. This type of calculation should prove very valuable in future work, especially when combined with the linearized extrapolations discussed in Sect.3.4.2 to reduce the labor involved in searching for the correct renormalized coupling constants.

Another approach has been taken very recently by BINDER [3.103,104], who considered the distribution function of the local order parameter in finite blocks. Using a real-space renormalization-group procedure based on the phenomenological scaling hypothesis, he has been able to obtain extremely encouraging results for the Ising model in two, three, and four dimensions (see also 3.108).

3.7 Conclusions

The combination of a renormalization-group analysis with Monte Carlo computer simulations has been demonstrated to provide a variety of powerful tools for the investigation of critical phenomena. There are several directions in which these methods can develop and some of them are easy to foresee. There will certainly be an increased application to more complicated and more realistic system now that the basic principles have been established. The properties of continuous variables and n-dimensional vector spins will be investigated with increasing accuracy, including the three-dimensional fcc and bcc antiferromagnets which have a large number of components in the order parameter. Phase diagrams including multicritical points will be calculated. Critical amplitudes will be calculated using MC simulations away from the critical point, possibly using series expansions with the derivations in the coefficients calculated from the correlation functions as in (3.18,28,29).

Other developments are more difficult to foresee. Applications to systems with quenched impurities and spin glasses would be highly desirable but have so far not be achieved. Fluids have not been treated at all, since the lattice has been an essential part of the definition of block spins.

Possibly the most interesting outstanding problem concerns critical dynamics. Progress has been made in this field by MA [3.26,105] and TOBOCHNIK et al. [3.106], but a generally applicable, accurate procedure has not yet been devised.

References

3.1 Th. Niemeijer, J.M.J. van Leeuwen: In *Phase Transitions and Critical Phenomena*, Vol.6, ed. by C. Domb, M.S. Green (Academic, New York 1976)
3.2 D.J. Wallace, R.K.P. Zia: Rep. Prog. Phys. *41*, 1 (1978)
3.3 K.G. Wilson, J. Kogut: Phys. Rep. *12*C, 76 (1974)
3.4 K.G. Wilson: Rev. Mod. Phys. *47*, 773 (1975)
3.5 N. Metropolis, A.W. Rosenbluth, M.N. Rosenbluth, A.H. Teller, E. Teller: J. Chem. Phys. *21*, 1087 (1953)
3.6 H. Müller-Krumbhaar, K. Binder: J. Stat. Phys. *8*, 1 (1973)
3.7 K. Binder: Adv. Phys. *23*, 917 (1974)

3.8 K. Binder: In *Phase Transitions and Critical Phenomena*, Vol.5b, ed. by C. Domb, M.S. Green (Academic, New York 1976) p.1
3.9 K. Binder: In *Monte Carlo Methods in Statistical Physics*, ed. by K. Binder, Topics in Current Physics, Vol.7 (Springer, Berlin, Heidelberg, New York 1979) p.1
3.10 B. Widom: J. Chem. Phys. *433*, 3898 (1965)
3.11 L.P. Kadanoff: Physics *2*, 263 (1966)
3.12 M.E. Fisher, R.J. Burford: Phys. Rev. *156*, 583 (1967)
3.13 M.E. Fisher, A.E. Ferdinand: Phys. Rev. Lett. *19*, 169 (1967)
3.14 A.E. Ferdinand, M.E. Fisher: Phys. Rev. *185*, 832 (1969)
3.15 M.E. Fisher: In *Proc. of Enrico Fermi International School of Physics*, ed. by M.S. Green (Academic, New York 1971)
3.16 E.K. Riedel: Phys. Rev. Lett. *28*, 675 (1972)
3.17 D.P. Landau: Phys. Rev. B *13*, 2997 (1976)
3.18 D.P. Landau: Phys. Rev. B *14*, 255 (1976)
3.19 E.K. Riedel, F.J. Wegner: Phys. Rev. Lett. *29*, 349 (1972)
3.20 F.J. Wegner: Phys. Rev. B *5*, 4529 (1972)
3.21 F.J. Wegner, E.K. Riedel: Phys. Rev. B *7*, 248 (1973)
3.22 E.K. Riedel, F.J. Wegner: Phys. Rev. B *9*, 294 (1974)
3.23 J.M.J. van Leeuwen: Phys. Rev. Lett. *34*, 1056 (1976)
3.24 M.P. Nightingale: Phys. Lett. *59*A, 486 (1977)
3.25 W. Kinzel, J.M. Yeomans: J. Phys. A *14*, L163 (1981)
3.26 S.K. Ma: Phys. Rev. Lett. *37*, 461 (1976)
3.27 Z. Friedman, J. Felsteiner: Phys. Rev. B *15*, 5317 (1977)
3.28 A.L. Lewis: Phys. Rev. B *16*, 1249 (1977)
3.29 P.J. Reynolds, H.E. Stanley, W. Klein: J. Phys. A *11*, L199 (1978)
3.30 P.J. Reynolds, H.E. Stanley, W. Klein: Phys. Rev. B *21*, 1223 (1980)
3.31 R.H. Swendsen: Phys. Rev. Lett. *42*, 859 (1979)
3.32 R.H. Swendsen, S. Krinsky: Phys. Rev. Lett. *43*, 177 (1979)
3.33 H.W.J. Blöte, R.H. Swendsen: Phys. Rev. Lett. *43*, 799 (1979)
3.34 H.W.J. Blöte, R.H. Swendsen: Phys. Rev. B *20*, 2077 (1979)
3.35 R.H. Swendsen: Phys. Rev. B *20*, 2080 (1979)
3.36 C. Rebbi, R.H. Swendsen: Phys. Rev. B *21*, 4094 (1980)
3.37 H.W.J. Blöte, R.H. Swendsen: Phys. Rev. B *22*, 4481 (1980)
3.38 R.H. Swendsen: *Cargèse, Lectures Notes, 1980 Summer School on Phase Transitions*, ed. by J.C. Le Guillou, J. Zinn (Plenum, New York 1980)
3.39 M. Novotny, D.P. Landau, R.H. Swendsen: Preprint
3.40 K.G. Wilson, R.H. Swendsen: Unpublished
3.41 D.P. Landau, R.H. Swendsen: Phys. Rev. Lett. *46*, 1437 (1981)
3.42 P.D. Eschbach, D. Stauffer, H.J. Herrmann: Phys. Rev. B *23*, 422 (1980)
3.43 P.S. Sahni, J.R. Banavar: Preprint
3.44 F. Fuctio, E. Marinari, G. Parisi, C. Rebbi: Preprint
3.45 H. De Raedt, A. Lagendijk: Phys. Rev. Lett. *46*, 77 (1981)
3.46 D.J. Scalapino, R.L. Sugar: Phys. Rev. Lett. *46*, 519 (1981)
3.47 M. Suzuki: Prog. Theor. Phys. *56*, 1454 (1976)
3.48 M. Barma, B. Shiram Shastry: Phys. Lett. *61*A, 15 (1977)
3.49 M. Barma, B. Shiram Shastry: Phys. Rev. B *18*, 3351 (1978)
3.50 M. Suzuki, S. Miyashita, A. Kuroda: Prog. Theor. Phys. *58*, 1377 (1977)
3.51 H.F. Trotter: Proc. Am. Math. Soc. *10*, 545 (1959)
3.52 M. Suzuki: Commun. Math. Phys. *51*, 183 (1976)
3.53 G. Shlifer, W. Klein, P.J. Reynolds, H.E. Stanley: J. Phys. A *12*, L169 (1979)
3.54 C.J. Lobb, K.A. Karsek: J. Phys. C *13*, L245 (1980)
3.55 T. Vicsek, J. Kertesz: J. Phys. A *14*, L31 (1981)
3.56 A. Baumgärtner: J. Phys. A *13*, L39 (1980)
3.57 A. Baumgärtner, K. Binder: Z. Phys. B *40*, 331 (1981)
3.58 K. Kremer, A. Baumgärtner, K. Binder: Z. für Phys. B*40*, 331 (1981)
3.59 S. Muto: Preprint
3.60 H.A. Kramers, G.H. Wannier: Phys. Rev. *60*, 252 (1941)
3.61 G.H. Wannier: Rev. Mod. Phys. *17*, 50 (1945)
3.62 L. Mittag, H.J. Steven: J. Math. Phys. (NY) *12*, 441 (1971)
3.63 F. Wegner: J. Math. Phys. (NY) *12*, 2259 (1971)

3.64 W. Greenberg: Commun. Math. Phys. *29*, 163 (1973)
3.65 F. Wegner: Physica *68*, 570 (1973)
3.66 F.Y. Wu, Y.K. Wang: J. Math. Phys. (NY) *17*, 439 (1976)
3.67 R. Fisch: J. Stat. Phys. *18*, 111 (1978)
3.68 R. Savit: Rev. Mod. Phys. *52*, 453 (1980)
3.69 I. Kondor: J. Phys. C *13*, L531 (1980)
3.70 W. Kinzel, E. Domany: Unpublished
3.71 H.R. Jauslin, R.H. Swendsen: Phys. Rev. B *24*, 313 (1981)
3.72 K.G. Wilson: Lectures at Les Houches, France, 1980
3.73 S.H. Shenker, J. Tobochnik: Phys. Rev. B, to be published, 1980
3.74 M. Creutz: Preprint
3.75 K.G. Wilson: Private communication
3.76 J.M.J. van Leeuwen: Phys. Rev. Lett. *34*, 1056 (1976)
3.77 F.Y. Wu: Phys. Rev. B *4*, 2312 (1971)
3.78 L.P. Kadanoff, F.J. Wegner: Phys. Rev. B *4*, 3989 (1971)
3.79 R.J. Baxter: Ann. Phys. (NY) *70*, 193 (1972)
3.80 N.N. Barber, R.J. Baxter: J. Phys. C *6*, 2913 (1973)
3.81 R.J. Baxter, S. Kelland: J. Phys. C *7*, L403 (1973)
3.82 S. Krinsky, R.H. Swendsen: Unpublished
3.83 R.B. Potts: Proc. Cambridge Philos. Soc. *48*, 106 (1952)
3.84 I.G. Enting: J. Phys. A *7*, 1617 (1964)
3.85 J.P. Straley, M.E. Fisher: J. Phys. A *6*, 1310 (1973)
3.86 R.J. Baxter: J. Phys. C *6*, L445 (1973)
3.87 D. Kim, R.I. Joseph: J. Phys. A *8*, 891 (1975)
3.88 A.N. Berker, S. Ostlund, F.A. Putnam: Phys. Rev. B *17*, 3650 (1978)
3.89 B. Nienhuis, A.N. Berker, E.K. Riedel, M. Schick: Phys. Rev. Lett. *43*, 737 (1979)
3.90 M.P.M. den Nijs: J. Phys. A *12*, 1857 (1979)
3.91 M. Nauenberg, D.J. Scalapino: Phys. Rev. Lett. *44*, 837 (1980)
3.92 J.L. Cardy, M. Nauenberg, D.J. Scalapino: Phys. Rev. B *22*, 2560 (1980)
3.93 R.B. Pearson: Phys. Rev. B *22*, 2579 (1980)
3.94 B. Nienhuis, E.K. Riedel, M. Schick: J. Phys. A *13*, L31 (1980)
3.95 B. Nienhuis, E.K. Riedel, M. Schick: J. Phys. A *13*, L189 (1980)
3.96 J.L. Black, V.J. Emery: Phys. Rev. B *23*, 429 (1980)
3.97 B. Nienhuis, M. Nauenberg: Phys. Rev. Lett. *35*, 477 (1975)
3.98 R.J. Baxter, F.Y. Wu: Phys. Rev. Lett. *31*, 1294 (1973)
3.99 Z. Racz, P. Rujan: Z. Phys. B *28*, 287 (1977)
3.100 W. Kinzel: Phys. Rev. B *19*, 4584 (1979)
3.101 H. Müller-Krumbhaar: Z. Phys. *35*, 339 (1979)
3.102 S. Muto, T. Oguchi, I. Ono: J. Phys. A *13*, 1799 (1980)
3.103 K. Binder: Z. für Phys. B*44*, 119 (1981)
3.104 K. Binder: Preprint
3.105 S.K. Ma: Phys. Rev. B *19*, 4824 (1979)
3.106 J. Tobochnik, S. Sarker, R. Cordery: Phys. Rev. Lett. *46*, 1417 (1981)
3.107 R.H. Swendsen, D. Andelman, A.N. Berter: Phys. Rev. B*24*, 6732 (1981)
3.108 R.H. Swendsen: Phys. Rev. Lett. *47*, 1159 (1981)

4. The Real Space Dynamic Renormalization Group

G. F. Mazenko and O. T. Valls

With 12 Figures

In this chapter the extension of real-space renormalization-group methods to dynamic phenomena is presented. Emphasis is placed on the calculation of space- and time-dependent correlation functions at a wide range of wave numbers, frequencies, and temperatures. The conditions that the renormalization-group transformation must satisfy in the dynamic case are elucidated. The systematic implementation of these conditions for the case of the square-lattice kinetic Ising model is given in detail, and results for the static and dynamic correlation functions in that case are presented.

4.1 Introduction

The renormalization group (RG) method [4.1] is a powerful tool in the theoretical study of condensed matter systems. If one is particularly interested in studying asymptotic critical properties in the scaling regime near a critical point, then the momentum space formulation [4.2] of the renormalization group is extremely useful. One of the reasons these methods work so well is that they build in the universality principle. Since nonuniversal and local details are irrelevant, they are essentially disregarded in these developments. The renormalization group approach (and its associated recursive analysis) is, however, not restricted to the study of universal properties near second-order phase transitions. From its inception [4.3] it has been understood that the real-space renormalization group (RSRG) method could be used to calculate thermodynamic quantities over the whole thermodynamic plane — not just near the critical point — and is capable of giving trustworthy information about the existence of first-order transitions. In particular one can calculate the magnetization and the specific heat [4.4] for the two-dimensional Ising model over the entire temperature range and obtain quantitatively reasonable results.

In this chapter we discuss the extension of the RSRG method to the treatment of dynamic phenomena. Most of the proposed schemes [4.5] for using RSRG methods to treat dynamic critical phenomena have focused on the calculation of the dynamic critical index z. As with the determination of the static critical indices, it is not clear how well these RSRG methods can compete with momentum space RG methods or high-temperature expansion methods in accuracy in treating the universal properties characterizing the critical point. We discuss this point in detail in [4.6].

As we pointed out above, the strength of the RSRG method is in treating the global behavior of a system. Consequently we will concentrate in this article on the application of RSRG ideas to the problem of the calculation of the space and time-dependent correlation functions typically encountered in experimental investigations of condensed matter systems. We will show that one can develop a systematic real-space dynamic renormalization group (RSDRG) [4.7-12] method for treating simple dynamical systems which is internally consistent, physically appealing, and quantitatively accurate. It also appears to have wide applicability. In order to keep the development as straightforward as possible, we will limit our discussion here to the simplest nontrivial case that we have studied thus far [4.10-12]. We will discuss the case of a set of spins on a two-dimensional square lattice with its statistical and dynamical properties governed by a single-spin-flip kinetic Ising model. Various other systems have been successfully studied, including the one-dimensional case with [4.13] and without [4.14] conserved spin. It is clear [4.15] that these methods can also be extended to treat strongly nonequilibrium situations.

In the next section we describe the kinetic Ising model we will study. In Sect.4.3 we develop the basic formal structure that is fundamental to the rest of our analysis. In Sect.4.4 we discuss the implementation of this formal development in the case where we break our Ising system up into a set of blocks of spins and treat the interaction between blocks in perturbation theory. An important aspect of the perturbation theory developed in Sect.4.4 is that we must use "effective" couplings [4.11.12] if we are to obtain good quantitative results. We discuss the determination of these effective couplings in Sect.4.5. This development involves more than just mathematics, but brings in several points of physics. In Sect.4.6 we present the results obtained from our theory for various observables like the magnetization, susceptibility, the static structure factor $\tilde{C}(q)$, and the dynamic structure factor $C(q,\omega)$. Finally, in Sect.4.7 we end with a brief discussion of future directions and unresolved questions.

4.2 Dynamic Problem of Interest

We consider a system of N Ising spins set on a lattice. The index i labels spins such that spin σ_i is located at position $\underline{r}_i$. Let σ stand for a given spin configuration $\sigma \equiv \{\sigma_1, \sigma_2, \ldots, \sigma_N\}$. The equilibrium probability distribution is given by

$$P[\sigma] = e^{H[\sigma]}/Z \quad , \tag{4.1}$$

where the partition function Z is the sum over all σ configurations

$$Z = \sum_{\sigma} e^{H[\sigma]} \quad , \tag{4.2}$$

$H[\sigma]$ is the Ising Hamiltonian [multiplied by $-\beta = -1/(k_B T)$]

$$H[\sigma] = \frac{K}{2} \sum_{i=1}^{N} \sum_{a} \sigma_i \sigma_{i+\delta_a} \quad , \tag{4.3}$$

K is the coupling constant ($K = +\beta J$ where J is the exchange interaction), and the sum over a is over basis vectors $\underline{\delta}_a$ connecting a spin at site i with its nearest neighbors. The equilibrium average of some quantity $A[\sigma]$ is defined as

$$\langle A[\sigma]\rangle = \sum_{\sigma} P[\sigma]\, A[\sigma] \quad . \tag{4.4}$$

It is conventional [4.16] in motivating the dynamic evolution of a set of Ising spins to picture the dynamics of the system as being driven by a heat reservoir. We assume that the resulting stochastic dynamics of our system is driven by a spin-flip operator (SFO)[1] $\tilde{D}_\sigma$ such that the equilibrium averaged time correlation function between some function $B[\sigma]$ of the σ spins at time zero and another function $A[\sigma]$ at time t is given by

$$C_{AB}(t) = \langle B[\sigma] e^{\tilde{D}_\sigma t} A[\sigma]\rangle \quad . \tag{4.5}$$

If $A[\sigma] = \sigma_i$ and $B[\sigma] = \sigma_j$, we have the usual spin-spin time correlation function, and if we take the space-time Fourier transform, we obtain the dynamic structure factor.

The SFO $\tilde{D}_\sigma$, which acts like a pseudo-Liouville operator, is a matrix which rotates one spin configuration σ' into another spin configuration σ:

$$\tilde{D}_\sigma A[\sigma] \equiv \sum_{\sigma'} \tilde{D}[\sigma|\sigma'] A[\sigma'] \quad . \tag{4.6}$$

The adjoint operator D_σ is defined by

$$D[\sigma|\sigma'] \equiv \tilde{D}[\sigma'|\sigma] \quad . \tag{4.7}$$

We will demand that $\tilde{D}_\sigma$ or equivalently D_σ satisfy two basic properties. First, we demand that

$$e^{D_\sigma t} P[\sigma] = P[\sigma] \tag{4.8}$$

or, equivalently,

$$\sum_{\sigma'} D[\sigma|\sigma'] P[\sigma'] = 0 \quad , \tag{4.9}$$

---

1 The general structure of these operators was discussed in [4.8,9] and, in particular, in [Ref.4.9, Appendix B].

which ensures the stationarity of the equilibrium probability distribution under time translations. A second condition follows if we remember that in a fully microscopic theory one has the symmetry relation

$$C_{AB}(t) = C_{BA}(-t) \quad . \tag{4.10}$$

If, under time reversal, the system is invariant and A and B have the same signature, then

$$C_{AB}(t) = C_{BA}(t) \quad . \tag{4.11}$$

We would like our stochastic dynamical model to satisfy this symmetry constraint. It is easy to see that (4.11) will be satisfied if the operator D_σ satisfies the detailed balance condition

$$D[\sigma|\sigma']P[\sigma'] = D[\sigma'|\sigma]P[\sigma] \quad . \tag{4.12}$$

Note that (4.7,12) give us the useful identity for any f:

$$\sum_{\sigma'} D[\sigma|\sigma']P[\sigma']f[\sigma'] = P[\sigma] \sum_{\sigma'} \tilde{D}[\sigma|\sigma']f[\sigma'] \quad . \tag{4.13}$$

Together, the stationarity condition (4.9) and the detailed balance condition (4.12) constrain the possible operators D_σ significantly. There is, however, still a great deal of latitude in choosing the appropriate operators, and operationally one is guided mainly by arguments of locality and simplicity. In this paper we will focus on a particular class of operators satisfying (4.9,12) where one flips a single spin on applying $\tilde{D}_\sigma$. This class of operators can be written in the form

$$\tilde{D}[\sigma|\sigma'] = -\frac{\alpha}{2} \sum_i \Lambda^{[i]}_{\sigma,\sigma'} \sigma_i \sigma_i' \, W_i[\sigma] \quad , \tag{4.14}$$

where α is the characteristic spin-flip-rate, the quantity

$$\Lambda^{[i]}_{\sigma,\sigma'} = \prod_{k(\neq i)} \delta_{\sigma_k,\sigma_k'} \quad , \tag{4.15}$$

guarantees that $\tilde{D}[\sigma|\sigma']$ is almost a diagonal matrix, and the probability of flipping, $W_i[\sigma]$ can be written in the form

$$W_i[\sigma] = e^{-\sigma_i E_i[\sigma]} \, V_i[\sigma] \tag{4.16}$$

where

$$E_i[\sigma] = K \sum_a \sigma_{i+\delta_a} \tag{4.17}$$

is the energy associated with the nearest neighbors of the spin at site i and $V_i[\sigma]$ is independent of σ_i.

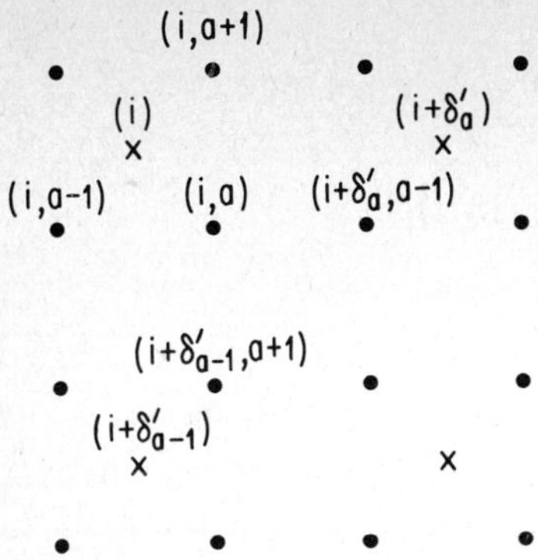

Fig.4.1. Specification of nearest-neighbor block (crosses) and σ spins (dots) with respect to a σ spin at site (i,a) on the square lattice

In this chapter we will focus specifically on the example of the square lattice. There are four vectors $\underline{\delta}_a$. We will treat the following form of $W_i[\sigma]$:

$$W_i[\sigma] = 1 + A\sigma_i \sum_a \sigma_{i+\delta_a} + A^2 \sum_a \sigma_{i+\delta_a}\sigma_{i+\delta_{a+1}} \tag{4.18}$$

where

$$A = -\frac{1}{2}\tanh 2K \quad . \tag{4.19}$$

We call this the "minimal coupling" form of $W_i[\sigma]$. In addition to its obvious locality and simplicity advantages, it has the property that its various terms are directly generated under renormalization [4.11]. After inserting (4.18) into (4.14) we see that $\tilde{D}_\sigma$ is completely specified. The goal of our theory is to calculate time correlation functions of the type given by (4.5) over a wide range of times and temperatures.

4.3 RSDRG - Formal Development

The renormalization group transformation function $T[\mu|\sigma]$ maps the set σ onto a new set $\mu \equiv \{\mu_1,\mu_2...\}$ of Ising spins located on a lattice geometrically similar to the initial one but having a larger lattice constant [4.3]. The Hamiltonian $H[\mu]$ governing the spins of the μ lattice is obtained from the Hamiltonian $H[\sigma]$ governing the σ spins via

$$e^{H[\mu]} = \sum_\sigma e^{H[\sigma]}T[\mu|\sigma] \quad . \tag{4.20}$$

The demand that the partition function be invariant under this transformation

$$Z = \sum_\sigma e^{H[\sigma]}$$

$$= \sum_\mu e^{H[\mu]} \tag{4.21}$$

requires that the transformation function satisfy the normalization condition

$$\sum_{\mu} T[\mu|\sigma] = 1 \quad . \tag{4.22}$$

Dividing (4.20) by Z, we find that the probability distribution governing the μ spins, $P[\mu]$, is given by

$$P[\mu] = \langle T[\mu|\sigma]\rangle \quad . \tag{4.23}$$

Notice then that (4.22) insures that the probability distribution is properly normalized. The coarse-graining function T presumably preserves the information about the long range and slowly varying degrees of freedom on the σ lattice by projecting out from the original probability distribution $P[\sigma]$ the slowly varying degrees of freedom and mapping them onto the new μ variables. If the original problem is described by a set of "coupling constants" K_i, the coarse-grained problem will be described by the new set K_i'. The recursion relation connecting these sets of coupling constants, $K_i' = K_i' (K_1, K_2, \ldots)$, can be used to investigate the scale invariance of the system and its behavior near a critical (scale invariant) point. This development is well known [4.3] and we will not dwell on it here. The point is that given such a recursion relation, one can go a very long way toward determining the thermodynamic behavior of a system.

We are primarily interested here in the structural and dynamic information contained in space and time-dependent correlation functions. Consider some function $A[\sigma]$ of the σ spins, for example, σ_i or $\sigma_i\sigma_{i+\delta_a}$. One can then use the physical interpretation of $T[\mu|\sigma]$ as a mapping function to identify the counterpart of $A[\sigma]$ on the μ lattice as

$$P[\mu]A[\mu] = \langle T[\mu|\sigma]A[\sigma]\rangle \quad . \tag{4.24}$$

We will refer to $A[\mu]$ as the "coarse-grained" variable associated with $A[\sigma]$. Notice, using (4.22) that the average of $A[\mu]$ over $P[\mu]$ equals the average of $A[\sigma]$ over $P[\sigma]$:

$$\langle A[\mu]\rangle' = \langle A[\sigma]\rangle \quad , \tag{4.25}$$

where $\langle\ \rangle'$ indicates an average over $P[\mu]$. Consequently the average of $A[\sigma]$ is preserved under the RSRG transformation with the proper identification of the coarse-grained variable. If we first use (4.20) to evaluate $P[\mu]$ in terms of the parameters characterizing $P[\sigma]$, use (4.24) to determine $A[\mu]$ explicitly in terms of μ, and insert this form for $A[\mu]$ into (4.25), we obtain recursion relations connecting correlation functions defined for the σ and μ lattices. We will give examples in Sect.4.4.

Within this coarse-graining procedure it is natural to introduce [4.11,14] a projection operator $\mathbb{P}$ which picks out of a σ-dependent quantity, $A[\sigma]$, the component "along" the block spin degrees of freedom. One can in general define this projection by the expression [4.12]

$$\mathbb{P}\, A[\sigma] \equiv \sum_{\mu} T[\mu|\sigma] \sum_{\mu'} \tilde{G}^{-1}[\mu|\mu'] \sum_{\sigma'} P[\sigma']T[\mu'|\sigma']A[\sigma'] \quad , \tag{4.26}$$

where

$$\tilde{G}[\mu|\mu'] = \langle T[\mu|\sigma]T[\mu'|\sigma]\rangle \quad , \tag{4.27}$$

$$\sum_{\mu''} \tilde{G}[\mu|\mu'']\tilde{G}^{-1}[\mu''|\mu'] = \delta_{\mu,\mu'} \quad , \tag{4.28}$$

$$\delta_{\mu,\mu'} = \prod_{i=1}^{N'} \delta_{\mu_i,\mu_i'} \quad , \tag{4.29}$$

and N' is the total number of cells. It is easy to see, using (4.28), that $\mathbb{P}$ is indeed a projection operator

$$\mathbb{P}^2 A[\sigma] = \mathbb{P}A[\sigma] \quad . \tag{4.30}$$

The interpretation of the projector $\mathbb{P}$ is complicated by the appearance of the matrix $\tilde{G}^{-1}$ in its definition. However, the meaning of $\mathbb{P}$ becomes rather transparent within the formalism for the correlation functions discussed below: it is then convenient, and it clarifies the interpretation of the coarse graining as a simple block spin transformation to require the normalization condition

$$\tilde{G}[\mu|\mu'] = \langle T[\mu|\sigma]T[\mu'|\sigma]\rangle = \delta_{\mu,\mu'}P[\mu] \tag{4.31}$$

to be satisfied. We have then

$$\tilde{G}^{-1}[\mu|\mu'] = \delta_{\mu,\mu'}\, P^{-1}[\mu] \quad , \tag{4.32}$$

and the projection operator becomes [4.10,11,14]

$$\begin{aligned} \mathbb{P}A[\sigma] &= \sum_{\mu} T[\mu|\sigma]P^{-1}[\mu] \sum_{\sigma'} P[\sigma']T[\mu|\sigma']A[\sigma'] \\ &= \sum_{\mu} T[\mu|\sigma]A[\mu] \quad . \end{aligned} \tag{4.33}$$

It is now obvious from the simple form (4.33) that the projector $\mathbb{P}$ indeed selects that part of a given variable $A[\sigma]$ which will be mapped onto the coarse-grained variable $A[\mu]$. Examples of the action of $\mathbb{P}$ on several variables are given in Sect.4.4.5, [see (4.112)].

It will be useful to note that $C_{AB}(t)$, defined by (4.5), can be rewritten in the form

$$C_{AB}(t) = \sum_{\sigma}\sum_{\sigma'} B[\sigma]A[\sigma']\, G[\sigma|\sigma';t] \quad , \tag{4.34}$$

where

$$G[\sigma|\sigma';t] = e^{D_\sigma t}\delta_{\sigma,\sigma'}P[\sigma'] \tag{4.35}$$

can be viewed as the generator of the time correlation functions on the σ lattice. We want to express $C_{AB}(t)$ in terms of the correlation functions defined with respect to the μ spins. As a first step in this analysis we can introduce the projection operator $\mathbb{P}$ and its complement

$$Q = \mathbb{I} - \mathbb{P} \tag{4.36}$$

into (4.5) to obtain

$$C_{AB}(t) = \sum_\sigma P[\sigma]\{(\mathbb{P} + Q)B[\sigma]\}\, e^{\tilde{D}_\sigma t}(\mathbb{P} + Q)A[\sigma] \quad . \tag{4.37}$$

The three terms in (4.37) involving Q's do not map onto the slow variables. These terms are to be thought of as being rapidly varying in space and time and amenable to direct perturbation theory analysis. In the term with two $\mathbb{P}$'s, the variables A and B are projected along their slow components, this being the term which we want to identify with time-dependent fluctuations on the μ lattice. We can rewrite this contribution as

$$\begin{aligned} C'_{AB}(t) &= \langle(\mathbb{P}\, B[\sigma])\; e^{\tilde{D}_\sigma t}(\mathbb{P}\, A[\sigma])\rangle \\ &= \sum_\mu \sum_{\mu'} \sum_\sigma P[\sigma]T[\mu'|\sigma]B[\mu']\; e^{\tilde{D}_\sigma t}T[\mu|\sigma]A[\mu] \\ &= \sum_\mu \sum_{\mu'} B[\mu']G[\mu|\mu';t]A[\mu] \quad , \end{aligned} \tag{4.38}$$

and

$$G[\mu|\mu';t] = \langle T[\mu'|\sigma]\; e^{\tilde{D}_\sigma t}\; T[\mu|\sigma]\rangle \tag{4.39}$$

has the direct interpretation, in comparison with (4.35), as the generator of time correlation functions on the μ lattice.

We now come to the key point in our development. We must identify the spin-flip operator appropriate to the dynamics of the μ spins. This identification is more difficult than the determination of the coarse-grained Hamiltonian since it involves the mapping of an operator onto an operator. The structure of the new dynamics is obvious if $G[\mu|\mu';t]$ can be written in the same form as $G[\sigma|\sigma';t]$ given by (4.35), that is, if we define a new SFO D_μ via

$$e^{D_\mu t}\delta_{\mu,\mu'}P[\mu'] = \langle T[\mu'|\sigma]e^{\tilde{D}_\sigma t}T[\mu|\sigma]\rangle \quad . \tag{4.40}$$

In writing (4.40) we have made the assumption that $\tilde{D}_\mu$, like $\tilde{D}_\sigma$, is a time-independent stochastic operator. This assumption is clearly not valid for all transformations $T[\mu|\sigma]$. For an arbitrary $T[\mu|\sigma]$, we must replace D_μ in (4.40) with a time-dependent quantity. We have investigated [4. 8] the forms of $D_\mu(t)$ corresponding to simple choices of $T[\mu|\sigma]$ and found severe problems associated with the generation of long-range interactions in space and time in $D_\mu(t)$. In this case then,

$D_\mu(t)$ is an operator of a very different kind than D_σ.[2] These problems can be completely avoided, however. The mapping $D_\sigma \to D_\mu$ will preserve the Markoffian nature of D_σ if the mapping function $T[\mu|\sigma]$ satisfies the eigenvalue-like equation [4.8]

$$\tilde{D}_\sigma T[\mu|\sigma] = D_\mu T[\mu|\sigma] \tag{4.41}$$

and the normalization condition (4.31). The "eigenvalue" in (4.41) is just the new SFO controlling the dynamics on the μ lattice.

In the next section we will discuss the practical aspects of constructing a transformation function $T[\mu|\sigma]$ satisfying (4.22,31,41). Let us for now assume we can carry out this construction and look at the consequences.

It is important to establish that the operator D_μ is of the same general type as D_σ. We can show this quite generally by first multiplying (4.41) by $T[\mu'|\sigma]P[\sigma]$, summing over σ, and using (4.31) to obtain

$$\langle T[\mu'|\sigma]\tilde{D}_\sigma T[\mu|\sigma]\rangle = D[\mu|\mu']P[\mu'] \quad . \tag{4.42}$$

From (4.7,13) we can easily show that

$$\langle T[\mu'|\sigma]\tilde{D}_\sigma T[\mu|\sigma]\rangle = \langle T[\mu|\sigma]\tilde{D}_\sigma T[\mu'|\sigma]\rangle \quad , \tag{4.43}$$

so that

$$D[\mu|\mu']P[\mu'] = D[\mu'|\mu]P[\mu] \quad . \tag{4.44}$$

We also have, using (4.22),

$$\begin{aligned}\sum_{\mu'} D[\mu|\mu']P[\mu'] &= \sum_\sigma P[\sigma]\tilde{D}_\sigma T[\mu|\sigma] \\ &= \sum_\sigma T[\mu|\sigma]D_\sigma P[\sigma] = 0 \quad .\end{aligned} \tag{4.45}$$

We see that both conditions (4.9,12) are satisfied by D_μ and that D_μ qualifies as a legitimate SFO.

Next we note, on using (4.40) in (4.38) that

$$\begin{aligned}C'_{AB}(t) &- \sum_\mu \sum_{\mu'} B[\mu']A[\mu]e^{D_\mu t}\,\delta_{\mu,\mu'}P[\mu'] \\ &= \sum_\mu \sum_{\mu'} B[\mu']\delta_{\mu,\mu'}P[\mu']e^{\tilde{D}_\mu t}A[\mu] \\ &= \langle B[\mu]e^{\tilde{D}_\mu t}A[\mu]\rangle' \quad ,\end{aligned} \tag{4.46}$$

and we can interpret $C'_{AB}(t)$ as the coarse-grained version of $C_{AB}(t)$. Returning to (4.37) we see that the use of (4.43) leads to further simplification. We easily

---

2 This basic question concerning the structure of the SFO generated after renormalization was essentially ignored in the RSDRG approaches in [4.5].

see that one of the cross terms can be written as

$$<(QB[\sigma])e^{\tilde{D}_\sigma t}(\mathbf{P}\,A[\sigma])>$$

$$= \sum_\mu A[\mu]<(QB[\sigma])e^{\tilde{D}_\sigma t}T[\mu|\sigma]>$$

$$= \sum_\mu A[\mu]e^{\tilde{D}_\mu t}<(QB[\sigma])T[\mu|\sigma]> \quad . \tag{4.47}$$

The last factor in (4.47) involves the coarse-grained variable associated with $QB[\sigma]$ which clearly must vanish. The cross terms in (4.37) therefore vanish and we have

$$C_{AB}(t) = C'_{AB}(t) + \bar{C}_{AB}(t) \quad , \tag{4.48}$$

where

$$\bar{C}_{AB}(t) = <(QB[\sigma])e^{\tilde{D}_\sigma t}(QA[\sigma])> \tag{4.49}$$

represents the correlation of short-range and rapidly varying degrees of freedom which we assume are amenable to a perturbative treatment. The long-range and slowly varying degrees of freedom which are represented by $C'_{AB}(t)$ are to be treated iteratively.

The recursion relation (4.48) is one of the fundamental results of this work. We note that the transformation $T[\mu|\sigma]$ implicit in (4.48) must satisfy not only the usual normalization condition (4.22) but also the eigenvalue equation (4.41) and the normalization condition (4.31).

4.4 Implementation of the RSDRG Using Perturbation Theory

4.4.1 General Development

We show in this section how the RSDRG formalism can be implemented systematically using a perturbation theory expansion. Let us assume that the Hamiltonian and SFO can be written as expansions in some ordering parameter ε:

$$H[\sigma] = \sum_{n=0}^{\infty} \varepsilon^n H^{(n)}[\sigma] \tag{4.50}$$

and

$$\tilde{D}_\sigma = \sum_{n=0} \varepsilon^n \tilde{D}_\sigma^{(n)} \quad . \tag{4.51}$$

We further assume that $H^0[\sigma]$ and $\tilde{D}_\sigma^0$ give a good zeroth-order description of the problem simple enough so that we can solve the zeroth-order problem completely. We must also be sure that the stationarity and detailed balance conditions are maintained order by order in ε.

The first step in the analysis is to solve the zeroth-order eigenvalue equation

$$\tilde{D}^0_\sigma T_0[\mu|\sigma] = D^0_\mu T_0[\mu|\sigma] \qquad (4.52)$$

subject to the normalization condition

$$\langle T_0[\mu|\sigma] T_0[\mu'|\sigma]\rangle_0 = \delta_{\mu,\mu'} P_0[\mu] \quad . \qquad (4.53)$$

In (4.53) $\langle\ \rangle_0$ denotes the average over the zeroth-order probability distribution

$$P_0[\sigma] = e^{H^{(0)}[\sigma]} / \sum_\sigma e^{H^{(0)}[\sigma]} \quad , \qquad (4.54)$$

and $P_0[\mu]$ is the zeroth-order probability distribution for the block spins whose dynamics, at this order, is given by the SFO $\tilde{D}^0_\mu$.

After the zeroth-order problem is solved, we can construct the general T, satisfying the eigenvalue equation (4.41) and the conditions (4.22,31), in two steps. First we construct a solution $\bar{T}[\mu|\sigma]$ to the eigenvalue equation

$$\tilde{D}_\sigma \bar{T}[\mu|\sigma] = \bar{D}_\mu \bar{T}[\mu|\sigma] \qquad (4.55)$$

that satisfies condition (4.22) and

$$\langle \bar{T}[\mu|\sigma] T_0[\mu|\sigma]\rangle_0 = \delta_{\mu,\mu'} P_0[\mu] \quad . \qquad (4.56)$$

We solve (4.55) perturbatively by assuming the expansions,

$$\bar{T}[\mu|\sigma] = \sum_{n=0}^{\infty} \varepsilon^n \bar{T}_n[\mu|\sigma] \qquad (4.57)$$

$$\bar{D}_\mu = \sum_{n=0}^{\infty} \varepsilon^n \bar{D}^{(n)}_\mu \qquad (4.58)$$

which we insert, along with (4.51), into (4.55) and equate the coefficients of like powers of ε. If we use the normalization (4.56), we then obtain, for example, the equations determining the first-order contributions to $\bar{D}_\mu$ and $\bar{T}[\mu|\sigma]$:

$$\bar{D}^{(1)}[\mu|\mu'] P_0[\mu'] = \langle T_0[\mu|\sigma] \tilde{D}^{(1)}_\sigma T_0[\mu'|\sigma]\rangle_0 \qquad (4.59)$$

and

$$\tilde{D}^0_\sigma \bar{T}_1[\mu|\sigma] - D^0_\mu \bar{T}_1[\mu|\sigma] = \bar{D}^{(1)}_\mu T_0[\mu|\sigma] - \tilde{D}^{(1)}_\sigma T_0[\mu|\sigma] \quad . \qquad (4.60)$$

We can continue on to systematically construct $\bar{D}^{(2)}$, $\bar{T}_2$, etc. [4.8,14,17]. Having constructed $\bar{T}[\mu|\sigma]$ in the form (4.57), we can then compute, using (4.50),

$$\bar{G}[\mu|\mu'] = \langle \bar{T}[\mu|\sigma] \bar{T}[\mu'|\sigma]\rangle \qquad (4.61)$$

to any desired order in ε. It may turn out that at a given order, $\bar{G}[\mu|\mu']$ is not diagonal. Then $\bar{T}$ does not satisfy the condition (4.31). However, one can always diagonalize $\bar{G}[\mu|\mu']$ through a rotation in μ space. We write (summation over repeated barred indices is implied)

$$T[\mu|\sigma] = S[\mu|\bar{\mu}]\bar{T}[\bar{\mu}|\sigma] \tag{4.62}$$

with S constructed at each order in perturbation theory so that (4.31) is satisfied,

$$S[\mu|\bar{\mu}]S[\mu'|\bar{\mu}']\bar{G}[\bar{\mu}|\bar{\mu}'] = \delta_{\mu,\mu'}P[\mu] \quad , \tag{4.63}$$

and the normalization (4.22) is maintained. It is then easily seen from (4.52) and (4.62) that $T[\mu|\sigma]$ is a solution of (4.41) with

$$D[\mu|\mu'] = S[\mu|\bar{\mu}]\bar{D}[\bar{\mu}|\bar{\mu}']S^{-1}[\bar{\mu}'|\mu'] \quad . \tag{4.64}$$

The main results of these calculations will be the renormalized probability distribution $P[\mu]$, the new SFO $\tilde{D}_\mu$, and the mapping function $T[\mu|\sigma]$. Assuming $P[\mu]$ and $\tilde{D}_\mu$ are of the same form as $P[\sigma]$ and $\tilde{D}_\sigma$ (that the RG procedure can be closed at a given order in ε), we can extract the recursion relations for the static and dynamic parameters characterizing $H[\sigma]$ and $\tilde{D}_\sigma$. Our main goal is to derive recursion relations of the form (4.48) relating correlation functions defined on different spatial scales. This derivation requires us to construct the coarse-grained variable $A[\mu]$ associated with the variable $A[\sigma]$. Since we now know $P[\sigma]$, $P[\mu]$, and $T[\mu|\sigma]$ to a given order in ε, we can use (4.24) to construct $A[\mu]$, $QA[\sigma]$, and $\bar{C}_{AB}(t)$ to the same order in ε. Finally, in (4.48), we insert in $C'_{AB}(t)$, given by (4.46), the explicit expression for $A[\mu]$ as a power series in ε and interpret the results as averages over the full probability distribution $P[\mu]$. We thus obtain a set of recursion relations that we can solve iteratively to give us the observables of interest.

4.4.2 Expansion for H and D_σ

We now implement this program explicitly for the case of the two-dimensional ferromagnet ($K > 0$) on a square lattice. This involves identification of the expansions for $H[\sigma]$ and $\tilde{D}_\sigma$ as in (4.50,51). As a first step we divide the system into cells and relabel the sites by assigning to each cell an index i and to a site within a cell the index a = 1,2,3,4 (Fig.4.1). We also introduce primitive lattice vectors δ'_a on the cell lattice. We can then rewrite the Hamiltonian (4.3) in the form

$$H[\sigma] = K_c V_0[\sigma] + K_I V_I[\sigma] \quad , \tag{4.65}$$

where

$$V_0[\sigma] = \sum_{i,a} \sigma_{i,a}\sigma_{i,a+1} \quad , \tag{4.66}$$

$$V_I[\sigma] = \frac{1}{2}\sum_{i,a} \sigma_{i,a}[\sigma_{i+\delta'_a,a-1} + \sigma_{i+\delta'_{a-1},a+1}] \quad . \tag{4.67}$$

Of course, before we artificially divide the lattice into cells, we have $K_c = K_I = K$. Our goal is to arrange matters so that the zeroth-order Hamiltonian, corresponding to uncoupled cells, gives a good description of the system. We hope then that the remaining interaction between cells is in terms of an "effective" coupling which we can treat as a small parameter. The identification of a good zeroth-order approximation and the effective coupling between cells requires a few rearrangements.

A spin at site i directly feels nearest-neighbor bonds of strength 4K. After breaking up the system into cells, however, the same spin, when we ignore couplings between cells, feels bonds of strength $2K_c$. If $K_c = K$, the error in the interaction felt at site i is large. We expect that a much better approximation is that $K_c \approx 2K$. This basic physics indicates that we should arrange our expansion such that the zeroth-order value of the cell coupling is different from the bulk value of the coupling. Thus we let $K_c \to K_c(\varepsilon) + \varepsilon[K_c - K_c(\varepsilon)]$ where $K_c(\varepsilon)$ possesses a power series expansion in the ordering parameter ε. In a similar fashion we expect that the coupling between cells will also be modified. Since some of the intercell correlations are taken into account through the choice of $K_c \sim 2K$, the intercell interaction will be reduced. We write, therefore,

$$H[\sigma] = \{K_c(\varepsilon) + \varepsilon[K_c - K_c(\varepsilon)]\}V_0[\sigma] + \varepsilon\{K_I(\varepsilon) + \varepsilon[K_I - K_I(\varepsilon)]\}V_I[\sigma] \tag{4.68}$$

and develop a perturbation expansion in powers of ε. We assume that $K_c(\varepsilon)$ and $K_I(\varepsilon)$ possess expansions:

$$K_c(\varepsilon) = K_0 + \sum_{n=1}^{\infty} \varepsilon^n K_n^0 \tag{4.69a}$$

$$K_I(\varepsilon) = K_R^0 + \sum_{n=1}^{\infty} \varepsilon^n K_n^R \quad . \tag{4.69b}$$

The coefficients K_0, K_n^0, K_R^0, K_n^R are to be determined by the imposition of constraints on an order by order basis. We shall do so in such a way as to ensure rapid convergence. Writing $H[\sigma]$ in the form (4.50) we can identify

$$H^{(0)}[\sigma] = K_0 V_0[\sigma] \quad , \tag{4.70}$$

$$H^{(1)}[\sigma] = \delta K_1 V_0[\sigma] + K_R^0 V_I[\sigma] \quad , \tag{4.71}$$

and

$$\delta K_1 = K - K_0 + K_1^0 \tag{4.72}$$

In constructing the $\tilde{D}_\sigma^{(n)}$ in (4.51) we must first, as with $H[\sigma]$, generalize $\tilde{D}_\sigma$ given in (4.14,18,19) to the case where the intercell interaction K_I and the intracell interaction K_c are different. The generalization of the minimal coupling form (4.18) to the case $K_c \neq K_I$ is

$$W_{i,a}[\sigma] = 1 + \sigma_{i,a}\Big[B_1(\sigma_{i,a+1} + \sigma_{i,a-1}) + B_2(\sigma_{i+\delta'_a,a-1} + \sigma_{i+\delta'_{a-1},a+1})\Big]$$
$$+ B_3\sigma_{i,a+1}\sigma_{i,a-1} + B_4(\sigma_{i,a+1}\sigma_{i+\delta'_a,a-1} + \sigma_{i,a-1}\sigma_{i+\delta'_{a-1},a+1})$$
$$+ B_5\sigma_{i+\delta'_a,a-1}\sigma_{i+\delta'_{a-1},a+1} \tag{4.73}$$

where, after some algebra, (4.9,12) imply

$$B_1 = \frac{a_c}{2(a_c + a_I)}\frac{(2a_I^2 a_c - a_I - a_c)}{(1 - a_I a_c)} , \tag{4.74a}$$

$$B_2 = \frac{a_I}{2(a_c + a_I)}\frac{(2a_c^2 a_I - a_I - a_c)}{(1 - a_I a_c)} , \tag{4.74b}$$

$$B_3 = \frac{a_I a_c^3}{(a_I + a_c)^2}\frac{(1 - a_I^2)}{(1 - a_I a_c)} , \tag{4.74c}$$

$$B_4 = \frac{1}{2}\frac{a_I a_c(a_I^2 + a_c^2 - 2a_I^2 a_c^2}{(a_I + a_c)^2(1 - a_I a_c)} , \tag{4.74d}$$

$$B_5 = \frac{a_c a_I^3}{(a_I + a_c)^2}\frac{(1 - a_c^2)}{(1 - a_I a_c)} , \tag{4.74e}$$

with

$$a_c = \tanh 2K_c , \tag{4.75}$$
$$a_I = \tanh 2K_I . \tag{4.76}$$

For $K_c = K_I = K$ we recover the minimal coupling form (4.18) with $B_1 = B_2 = A$ and $B_3 = B_4 = B_5 = A^2$. If we again expand K_c and K_I as in (4.68,69), we obtain an expansion in powers of ε for W_i,a. In order to obtain $\tilde{D}_\sigma$ in an expansion in ε we must allow for an expansion for α of the same type as for K_c:

$$\alpha = \alpha_0(\varepsilon) + \varepsilon[\alpha - \alpha_0(\varepsilon)]$$
$$= \alpha_0 + \delta\alpha_1\varepsilon + O(\varepsilon^2) , \tag{4.77}$$

where

$$\delta\alpha_1 = \alpha_0^{(1)} + \alpha - \alpha_0 . \tag{4.78}$$

Combining these results we easily obtain the first two terms in (4.51):

$$\tilde{D}_\sigma^0[\sigma|\sigma'] = \sum_i \tilde{D}^{i,0}[\sigma|\sigma'] \tag{4.79}$$

$$\tilde{D}^{i,0}[\sigma|\sigma'] = -\frac{\alpha_0}{2}\sum_a \Lambda_{\sigma,\sigma'}^{[i,a]}\left[1 - \frac{a_0}{2}\sigma_{i,a}(\sigma_{i,a+1} + \sigma_{i,a-1})\right]\sigma_{i,a}\sigma'_{i,a} \tag{4.80}$$

where

$$a_0 = \tanh 2K_0 \quad , \tag{4.81}$$

and

$$\tilde{D}^{(1)}[\sigma|\sigma'] = -\frac{\alpha_0}{2}\sum_{i,a} \Lambda^{[i,a]}_{\sigma,\sigma'} W^{(1)}_{i,a}[\sigma]\, \sigma_{i,a}\sigma'_{i,a} + \frac{\delta\alpha_1}{\alpha_0}\tilde{D}^0[\sigma|\sigma'] \tag{4.82}$$

where $W^{(1)}_{i,a}[\sigma|\sigma']$ is obtained by expanding the B_i's given by (4.74) to $O(\varepsilon)$ and then reading off the associated first-order contributions from (4.73). All of the terms but B_5 contribute to $O(\varepsilon)$.

4.4.3 Solution to the Zeroth-Order Problem

The first step in solving the zeroth-order problem is to identify the fast and slow degrees of freedom in a cell. This identification follows from the solution of the eigenvalue equation

$$\tilde{D}^{i,0}_{\sigma}\psi^{(n)}_i(\sigma) = -\lambda^{(n)}\psi^{(n)}_i(\sigma) \quad . \tag{4.83}$$

The operator $\tilde{D}^{i,0}_{\sigma}$ is given by (4.80). In the case of 4 spins in a cell there are 16 eigenfunctions which can be properly orthogonalized,

$$\langle\psi^{(n)}_i(\sigma)\psi^{(n')}_i(\sigma)\rangle_0 = \delta_{n,n'} \quad , \tag{4.84}$$

and form a complete set,

$$\sum_n \psi^{(n)}_i(\sigma)\psi^{(n)}_i(\sigma')^* P_0[\sigma'] = \delta^i_{\sigma,\sigma'} \tag{4.85}$$

where $\delta^i_{\sigma,\sigma'} = \Pi^4_{a=1}\,\delta_{\sigma_{i,a}\sigma'_{i,a}}$. Since $\tilde{D}^{i,0}_{\sigma}$ is invariant under the flipping of all four spins in the cell, the eigenfunctions divide up into eight with even symmetry and eight with odd symmetry under the flipping of the four spins. The explicit forms for the $\psi^{(n)}_i$ and $\lambda^{(n)}$ are given in [4.12]. For our purposes it is sufficient to note that for couplings $K > 0$ there are two modes which vary more slowly (have a smaller $\lambda^{(n)}$) than the others. These modes are listed in Table 4.1. The normalization of the $\psi^{(n)}_i$ is determined by (4. 84). In particular

$$N_1^2 = [4(1 + 2r + s)]^{-1} \tag{4.86}$$

where r and s are the zeroth-order nearest- and next-nearest neighbor correlation functions given by

$$r = \langle\sigma_{i,a}\sigma_{i,a\pm1}\rangle_0 = \frac{a_0}{2 - a_0^2} \quad , \tag{4.87}$$

Table 4.1. Slow eigenfunctions and eigenvalues at zeroth order

n	$\psi_i^{(n)}(\sigma)$	$\lambda^{(n)}$
0	1	0
1	$N_1 \sum_a \sigma_{i,a}$	$\alpha_0(1 - a_0)$

$$s = \langle\sigma_{i,a}\sigma_{i,a\pm2}\rangle_0 = \frac{a_0^2}{2 - a_0^2} \; . \tag{4.00}$$

The zeroth-order mapping function satisfying (4.52) is given by

$$T_0[\mu|\sigma] = \prod_{i=1}^{N/4} T_0^i[\mu|\sigma] \tag{4.89}$$

where

$$T_0^i[\mu|\sigma] = \frac{1}{2}\left[1 + \mu_i\psi_i^{(1)}(\sigma)\right] \tag{4.90}$$

and the zeroth order renormalized SFO is

$$D^0[\mu|\mu'] = -\frac{\lambda^{(1)}}{2}\sum_i \Lambda_{\mu,\mu'}^{[i]}\,\mu_i\mu_i' \; . \tag{4.91}$$

We easily identify at zeroth order the renormalized flipping rate as

$$\alpha' = \lambda^{(1)} \; . \tag{4.92}$$

It is straightforward, using the orthogonality condition (4.84), to show that $T_0[\mu|\sigma]$ satisfies the normalization (4.53) with the trivial zeroth result for the coarse-grained probability distribution:

$$P_0[\mu] = (1/2)^{N/4} \; . \tag{4.93}$$

Note, in principle, that we could choose any of the eigenfunctions $\psi_i^{(n)}$ in place of $\psi_i^{(1)}$ and still satisfy (4.52,53). The physics of the situation where we interpret T_0 as selecting out the slowest modes dictates that we choose $\psi_i^{(1)}$.

Without any reference to the dynamics, in transformations of the form given by (4.89,90), the functions ψ_i, which respect symmetry under interchange of spins, are of the general form,

$$\psi_i = N_1 \sum_a (\sigma_{i,a} + f\sigma_{i,a+1}\sigma_{i,a-1}\sigma_{i,a}) \; , \tag{4.94}$$

where f is arbitrary. The function $\psi_i^{(1)}$ corresponds, then, to the choice f = 0 to which one might be led merely from simplicity considerations and the fact that the total spin is the order parameter in this problem.[3] Other static RG calculations [4.4b] have chosen N_1 and f independently through the introduction of various constraint conditions.

4.4.4 Renormalization to First Order

The first step in our first order analysis is the evaluation of $\bar{G}[\mu|\mu']$ defined by (4.61) to $O(\varepsilon)$. In order to simplify matters a bit and since this choice will not influence any of our numerical results in Sect.4.6 we set $\delta K_1 = 0$ by choosing K_1^0 appropriately. We easily find then, after noting that we do not need $\bar{T}_1[\mu|\sigma]$ to $O(\varepsilon)$ due to the normalization condition (4.53), that

$$\bar{G}[\mu|\mu'] = \delta_{\mu,\mu'} P[\mu] \quad , \tag{4.95}$$

where

$$P[\mu] = P_0[\mu]\left\{1 + 2\varepsilon\nu_1^2 K_R^0 H_1[\mu] + O(\varepsilon^2)\right\} \tag{4.96}$$

$$\nu_1 = \langle\sigma_{i,a}\psi_i^{(1)}(\sigma)\rangle_0 = \frac{1}{2}\,(1 + 2r + s)^{\frac{1}{2}} \tag{4.97}$$

and

$$H_1[\mu] = \frac{1}{2}\sum_{i,a} \mu_i\mu_{i+\delta_a'} \quad . \tag{4.98}$$

Since $\bar{G}$ is diagonal in μ and μ', no rotation is needed and $T[\mu|\sigma] = \bar{T}[\mu|\sigma] + O(\varepsilon^2)$. Next we see on exponentiating the $O(\varepsilon)$ term in $P[\mu]$ that we obtain a renormalized nearest neighbor Ising Hamiltonian with coupling

$$K' = 2\nu_1^2 K_R^0 \quad . \tag{4.99}$$

Moving to the dynamic renormalization, $\bar{D}_\mu^{(1)}$ is obtained by taking the zeroth-order average indicated by (4.59) where $\tilde{D}_\sigma^{(1)}$ is given by (4.82). We easily find

$$\bar{D}^{(1)}[\mu|\mu'] = \sum_i \Lambda_{\mu,\mu'}^{[i]} \frac{\mu_i\mu_i'}{2}\left[\delta\lambda^{(1)} - \lambda^{(1)}K'\mu_i' \sum_a \mu_{i+\delta_a'}'\right] \quad , \tag{4.100}$$

where K' is given by (4.99) and

$$\delta\lambda^{(1)} = \delta\alpha_1 \quad . \tag{4.101}$$

Since there is no rotation to this order, we have $D_\mu^{(1)} = \bar{D}_\mu^{(1)}$. After adding $D_\mu^{(1)}$ to D_μ^0, we obtain a $\tilde{D}_\mu$ of the same form as $\tilde{D}_\sigma$ (expanded to first order in K) but with the new coupling K' given by (4.99) and the new flipping rate

---

3 In studying the case of an anti-ferromagnetic Ising model on a square lattice a direct application of our procedures leads us to choose $\psi_i \sim \sum_a(-1)^a\,\sigma_{i,a}$. Again this is the slowest odd mode and the local order parameter.

$$\alpha' = \lambda^{(1)} + \varepsilon\delta\alpha_1 + O(\varepsilon^2) \quad . \tag{4.102}$$

The quantity $\delta\alpha_1$ can be chosen to vanish. We have then

$$\alpha' = \lambda^{(1)} + O(\varepsilon^2) \quad . \tag{4.103}$$

Before renormalization, space and time scales are set by the lattice spacing a and the decay time α^{-1}. After renormalization we have the new lattice spacing $a' = ba$ and decay time $(\alpha')^{-1}$. In this sense the original and renormalized problems are different even at the fixed point where $K' = K$. We can however rescale lengths via $L \to L' = L/b$ and times $t \to t' = t(\alpha'/\alpha)$ so that, for example, $\tilde{D}_\mu t = (\alpha/\alpha')\tilde{D}_\mu t'$. Then at the critical point, for an infinite system, $(\alpha/\alpha')\tilde{D}_\mu$ is identical in structure to $\tilde{D}_\sigma$ and approaches a finite fixed point form. This time rescaling is essentially a statement of dynamic scaling. It is conventional to express the time rescaling in the form

$$t' = t/b^z \quad , \tag{4.104}$$

where z is the dynamic critical index, which would be equal to one if space and time rescaled in the same way. If we write

$$\alpha'/\alpha \equiv \Delta = \lambda^{(1)}/\alpha \quad , \tag{4.105}$$

then from (4.104,105) with $b = 2$ we have

$$z = -\ln\Delta/\ln 2 \quad . \tag{4.106}$$

4.4.5 Recursion Relations for the Correlation Functions

We can now derive the recursion relations for the correlation functions of interest. The first step in the construction of these recursion relations is the determination of the various coarse-grained variables using (4.24). The coarse-grained variables corresponding to $\sigma_{i,a}$, $\sigma_{i,a}\sigma_{j,a'}$ $(i \neq j)$ and $\sigma_{i,a}\sigma_{i,a'}$, denoted by $M_{i,a}[\mu]$, $\Pi^I_{i,a;j,a'}[\mu]$, and $\Pi^C_{i,aa'}[\mu]$, respectively, are given to zeroth order by

$$M_{i,a}[\mu] = \nu_1\mu_i + O(\varepsilon) \quad , \tag{4.107a}$$

$$\prod^{I}_{i,a;j,a'}[\mu] = \nu_1^2\mu_i\mu_j + O(\varepsilon) \quad , \tag{4.107b}$$

$$\prod^{C}_{i,aa'}[\mu] = \delta_{a,a'} + r(\delta_{a',a+1} + \delta_{a',a-1}) + s\delta_{a',a+2} + O(\varepsilon) \quad . \tag{4.107c}$$

The lowest order recursion relation for the magnetization,[4]

4 A more careful analysis of the ordered phase and the associated spontaneous symmetry breaking is required than we give here. This question was treated in detail by G.F. Mazenko and O.T. Valls, Phys. Rev. B, in press.

$$m = \langle\sigma_{i,a}\rangle \quad , \tag{4.108}$$

is given by

$$\begin{aligned} m = \langle\sigma_{i,a}\rangle &= \sum P[\mu] M_{i,a}[\mu] \\ &= \sum_{\mu} P[\mu][\nu_1\mu_i + O(\varepsilon)] \\ &= \nu_1 m' + O(\varepsilon) \quad . \end{aligned} \tag{4.109}$$

We can then easily obtain recursion relations for static correlation functions of the form

$$\varepsilon(m,n) = \frac{1}{N}\sum_i \langle\sigma_i\sigma_{i+m\hat{x}+n\hat{y}}\rangle \tag{4.110}$$

m,n integers, by expressing $\sigma_i\sigma_{i+m\hat{x}}$ in terms of Π^I and Π^C and then using (4.25) to obtain, for example,

$$\varepsilon(1,0) = \frac{1}{2} r + \frac{1}{2}\nu_1^2\varepsilon'(1,0) \quad , \tag{4.111a}$$

$$\varepsilon(2,0) = \nu_1^2\varepsilon'(1,0) \quad , \tag{4.111b}$$

$$\varepsilon(1,1) = \frac{1}{4}[S + 2\varepsilon'(1,0) + \varepsilon'(1,1) \quad , \tag{4.111c}$$

$$\varepsilon(4,0) = \nu_1^2\varepsilon'(2,0) \quad . \tag{4.111d}$$

General formulae depend on whether m and n are even or odd and we will not write them down here.

We turn now to the space- and time-dependent correlation functions such as the dynamic structure factor which satisfy the general recursion relation (4.48). To obtain a recursion relation for the dynamic structure factor $C_{i,a;j,a'}(t)$ we choose, in (4.48), $A[\sigma] = \delta\sigma_{i,a}$ and $B[\sigma] = \delta\sigma_{j,a'}$, where $\delta\sigma_{i,a} = \sigma_{i,a} - \langle\sigma_{i,a}\rangle$. Clearly the collective variable associated with $\delta\sigma_{i,a}$ is, to lowest order in ε, $\nu_1(\mu_i - m')$. Evaluation of the term $\bar{C}_{AB}(t)$ in (4.48) to lowest order in ε requires determination of the "fast" part of $\sigma_{i,a}$. We find

$$\mathbb{P}\sigma_{i,a} = \nu_1\psi_i^{(1)}(\sigma) \quad , \tag{4.112a}$$

$$Q\sigma_{i,a} = \sum_n (1 - \delta_{n,1})\nu_n(a)\psi_i^{(n)}(\sigma) \quad , \tag{4.112b}$$

where

$$\nu_n(a) = \langle\sigma_{i,a}\psi_i^{(n)}(\sigma)\rangle_0 \tag{4.113}$$

and the orthogonality and completeness of the set of cell eigenfunctions has been used. We can then evaluate $\bar{C}_{i,a;j,a'}(t)$ directly to lowest order in ε by replacing $\tilde{D}_\sigma$ by $\tilde{D}^0_\sigma$, $P[\sigma]$ by $P_0[\sigma]$, and using the eigenvalue equation to diagonalize $e^{D^0_\sigma t}$. The result is

$$\bar{C}_{i,a;j,a'}(t) = \delta_{ij} \sum_n (1 - \delta_{n,1}) \nu_n^*(a) \nu_n(a') e^{-\lambda^{(n)} t} \quad . \tag{4.114}$$

Notice that this quantity is very local in space and rapidly decaying since it depends (note that $\nu_0(a) = 0$ due to symmetry) on the "fast" modes $\lambda^{(n)} > \lambda^{(1)}$. After using our results for the collective variable associated with $\delta\sigma_{i,a}$, we obtain from (4.48)

$$C_{i,a;j,a'}(t) = \nu_1^2 C'_{ij}(t') + \delta_{ij} \sum_n (1 - \delta_{n,1}) \nu_n^*(a) \nu_n(a') e^{-\lambda^{(n)} t} \tag{4.115}$$

where $C'_{i,j}(t')$ is the spin correlation function for the μ spins at $t' = \Delta t$, with dynamics governed by the SFO $\tilde{D}'_\mu = \tilde{D}_\mu/\Delta$ where Δ is given by (4.105). Thus, we have rescaled time so that $\tilde{D}_\sigma$ and $\tilde{D}'_\mu$ have the same time scale α^{-1}. If we introduce the Fourier-Laplace transform of $C_{ij}(t)$

$$C(q,Z) = \frac{1}{N} \sum_{i,j} e^{i\underline{q} \cdot (\underline{r}_i - \underline{r}_j)} (-i) \int_0^{+\infty} dt\, e^{iZt} C_{ij}(t) \quad , \tag{4.116}$$

then we find from (4.115) that

$$C(q,Z) = \frac{f(q)}{\Delta} C'(2q,Z') + \sum_n (1 - \delta_{n,1}) \frac{f_n(q)}{Z + i\lambda^{(n)}} \quad , \tag{4.117}$$

where

$$f_n(q) = \frac{1}{4} \sum_a \left[\nu_n^*(a)\nu_n(a) + 2\nu_n^*(a)\nu_n(a+1) g_1(q) + \nu_n^*(a)\nu_n(a+2) g_2(q) \right] \quad , \tag{4.118}$$

$$f(q) = 1 + \cos q_x + \cos q_y + \cos q_x \cos q_y \quad , \tag{4.119}$$

$$g_1(q) = \frac{1}{2} (\cos q_x + \cos q_y) \quad , \tag{4.120}$$

and

$$g_2(q) = \cos q_x \cos q_y \quad . \tag{4.121}$$

All wave numbers are measured in terms of the inverse lattice constant which is fixed at unity. Therefore they must be rescaled by a factor of two upon iteration and $Z' = Z/\Delta$. The recursion relation for the dynamic structure factor, $C(q,\omega)$, is obtained from (4.117) by the use of the relation

$$C(q,\omega) = -2\,\mathrm{Im}\{C(q, Z = \omega + i0^+)\} \quad . \tag{4.122}$$

The recursion relation for the static structure factor can be extracted from (4.117) through the use of the identity

$$\tilde{C}(q) = \lim_{Z\to\infty} ZC(q,Z) \quad . \tag{4.123}$$

We obtain, after using the completeness of the eigenfunctions $\psi^{(n)}$, that the static structure factor satisfies the recursion relation

$$\tilde{C}(q) = 1 + 2rg_1(q) + sg_2(q) - \nu_1^2 f(q) + \nu_1^2 f(q)\tilde{C}'(2q) \tag{4.124}$$

which, of course, can be obtained directly by setting $t = 0$ in (4.115) and Fourier transforming over space. The static susceptibility χ is simply the $q \to 0$ limit of $\tilde{C}(q)$ and satisfies the very simple recursion relation

$$X = 4\nu_1^2 X' \quad . \tag{4.125}$$

In deriving this result we made use of the identity $1 + 2r + s - 4\nu_1^2 = 0$. This treatment must be modified (see footnote 4) somewhat for $T < T_c$.

4.5 Determination of Parameters

4.5.1 General Comments

The recursion relations given, for example, by (4.117) are not yet completely specified. We note that the ν_{n_0} and $\lambda^{(n)}$ depend on K_0 and α_0, while K', as given by (4.99), depends on K_0 and K_R^0 and $\alpha' \equiv \lambda^{(1)} = \alpha_0(1 - a_0)$. Consequently we must determine K_0, K_R^0, and α_0. One might first simply set $K_0 = K_R^0 = K$ and $\alpha_0 = \alpha$. However, for reasons we discuss below and also from the arguments we made in introducing K_0 and K_R^0, this choice would not lead to reasonable results at lowest order.

4.5.2 The Parameters K_0 and K_R^0

Our basic method for the determination of K_0 and K_R^0 comes from an analysis of the space-dependent correlation functions of the form given by (4.111). Consider a correlation function between two spins separated by a distance R along the x axis. Consider first the high temperature phase. If we set $t = 0$ in (4.115), we find for $R \gg 1$

$$\langle\sigma_0\sigma_R\rangle = \nu_1^2\langle\mu_0\mu_{R/b}\rangle \quad , \tag{4.126}$$

where b is the length rescaling parameter ($b = 2$ here). With $R = b^n$ we find after n iterations

$$\langle\sigma_0\sigma_R\rangle \simeq \left\{\prod_{m=0}^{n-1} \nu_1^2[K^{(m)}]\right\}\langle\mu_0\mu_1\rangle_n \quad , \tag{4.127}$$

where μ_0 and μ_1 are nearest neighbors and $K^{(i)}$ the coupling after i iterations. After many iterations we flow to the high-temperature fixed point. For high temperatures we have $\langle\mu_0\mu_1\rangle_n \sim K^{(n)}$ and $\nu_1 \to 1/2$. Therefore, the first factor in (4.127) is $\approx (\nu_1^2(0))^n \simeq R^{-\Gamma}$, $\Gamma = 2|\ln\nu_1(0)|/\ln b$, and gives algebraic decay with distance. To obtain exponential decay we require $K^{(n)} \propto (K)^R = (K)^{b^n}$. Therefore, near the high-temperature fixed point, the recursion relation should be of the form

$$K^{(n+1)} \propto [K^{(n)}]^b \quad . \tag{4.128}$$

From our recursion relation (4.99) we have, as $K \to 0$, $K' = K_R^0/2$. If $K_R^0 \propto K$ is choosen for small K, the unphysical result that correlation functions decay algebraically in the disordered phase is obtained.

A similar argument shows that the limiting form of the recursion relation at low temperature should be

$$K^{(n+1)} = bK^{(n)} \quad . \tag{4.129}$$

Acceptable recursion relations must interpolate smoothly between the behaviors (4.128,129).

We want to combine this information with our earlier comments about bond cutting and the choice of the cell coupling K_0. We will address these questions through an analysis of the short-range correlation functions of the form given by (4.111) for small m or n. We formulate the theory in terms of these quantities because we should know a good deal about them in any problem of interest. In the present case we know them exactly, while in other siutations we may need to analyze them using a high- and low-temperature analysis or Monte Carlo simulations. It is not difficult to see that we can construct a simple relation between K' and K that satisfies both (4.128,129) and which is independent of K_0 by taking the ratios of the recursion relations for $\varepsilon(4,0)$ and $\varepsilon(2,0)$:

$$\frac{\varepsilon(4,0)}{\varepsilon(2,0)} = \frac{\varepsilon'(2,0)}{\varepsilon'(1,0)} \quad . \tag{4.130}$$

A numerical solution of this equation to give $K' = K'(K)$ is rather involved. We have found [4.10,11] that a rather good approximate solution to (4.130) is given by

$$\phi' = \phi^2 \tag{4.131}$$

with $\phi = \exp(2K)\tanh(K)$. This recursion relation has several remarkable properties: (a) it satisfies (4.128,129); (b) its fixed point solution $\phi^* = 1$ gives the exact transition temperature $\tanh K = \sqrt{2} - 1$; and (c) since $(\partial K'/\partial K)^* = (\partial\phi'/\partial\phi)^* = 2$, the associated thermal index $y_T = \ln(\partial K'/\partial K)^*/\ln 2 = 1$ agrees with the exact value. For these reasons we have chosen to work with the recursion relation (4.131) in the rest of this chapter. Once $K'(K)$ is found, K_R^0 is determined by (4.99).

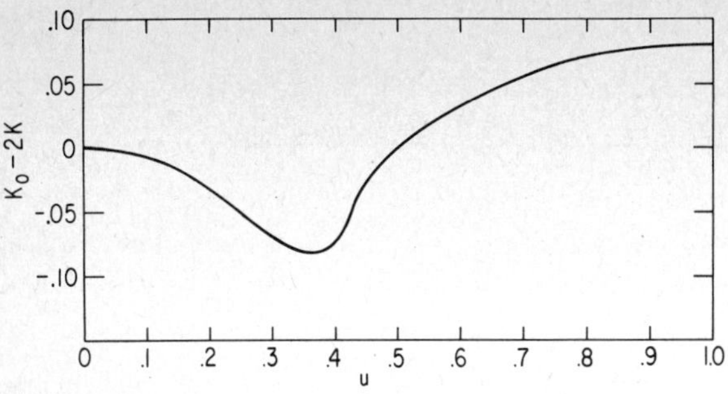

Fig.4.2. The difference between the intracell bond K_0 and double the lattice bond K as a function of u = tanh K

Next, we specify K_0. The nearest-neighbor recursion relation (4.111a) yields a value of a_0 (r and ν_1^2 depend on a_0 only) as a function of $\varepsilon(1,0)$ and $\varepsilon'(1,0)$. In the present case $\varepsilon(1,0)$ is known exactly. This knowledge is not crucial: the result obtained for K_0 is quite similar whether exact expressions or approximate series expansions are used for $\varepsilon(1,0)$. In Fig.4.2, we show that K_0 - 2K is small over the whole temperature range, as expected from our discussion of bond cutting.

4.5.3 The Dynamic Parameters α_0 and Δ

From our experience with K_0, we would not expect the approximation $\alpha_0 = \alpha$ to work very well. This is confirmed in the treatment of the one-dimensional kinetic Ising model [4.14] where the approximation $\alpha_0 = \alpha$ leads to results which, at zeroth order, badly disagree with the exact solutions which exist in that case. As a first step in developing a quantitatively accurate method for determining Δ [and therefore $\alpha_0 = \Delta\alpha/(1 - a_0)$] we take the small q and Z limits of (4.115) to obtain

$$C(0,0) = \frac{4\nu_1^2}{\Delta} C'(0,0) \tag{4.132}$$

or

$$\Delta = 4\nu_1^2 C'(0,0)/C(0,0) \quad . \tag{4.133}$$

Using standard procedures [4.6] it is not difficult to show that

$$C(q,Z) = \langle\sigma_{-q}R(Z)\sigma_q\rangle \tag{4.134}$$

[where σ_q is the spatial Fourier transform of σ_i and R(Z) is the resolvent operator: $R(Z) = (Z - i\tilde{D}_\sigma)^{-1}$] can be written in the form

$$C(q,Z) = \tilde{C}(q)[Z + i\phi(q,Z)]^{-1} \tag{4.135}$$

where the memory function $\phi(q,Z)$ is the sum of static and dynamic parts $\phi(q,Z) = \phi^{(s)}(q) + \phi^{(d)}(q,Z)$, with

$$\varphi^{(s)}(q)\tilde{C}(q) = - \langle\sigma_{-q}\tilde{D}_\sigma\sigma_q\rangle \quad , \tag{4.136a}$$

$$\varphi^{(d)}(q,Z)\tilde{C}(q) = -i\Big\{\langle(\tilde{D}_\sigma\sigma_{-q})R(Z)(\tilde{D}_\sigma\sigma_q)\rangle - \langle(\tilde{D}_\sigma\sigma_{-q})R(Z)\sigma_q\rangle$$

$$\times\, C^{-1}(q,Z)\, \langle\sigma_{-q}R(Z)(\tilde{D}_{\sigma}\sigma_{q})\rangle\Big\} \quad . \tag{4.136b}$$

If we insert the form given by (4.135) for C and C' into (4.132), and use the result (4.125), we obtain the simple result:

$$\Delta = \frac{\phi(0,0)}{\phi'(0,0)} \tag{4.137}$$

(the rescaling of D_μ discussed below (4.115) must be kept in mind). As we shall indicate below we can evaluate the right-hand side of (4.136a) directly in terms of known static correlation functions. It is therefore convenient to write

$$\phi(0,0) = \phi^{(s)}(0)(1 - R) \tag{4.138}$$

with

$$R = \lim_{Z\to i0+}\ \lim_{q\to 0} - \frac{\phi^{(d)}(q,Z)}{\phi^{(s)}(q)} \quad . \tag{4.139}$$

We can rewrite (4.137) in the form

$$\Delta = \Delta_0 Q \tag{4.140}$$

where

$$\Delta_0 = \frac{\phi^{(s)}(0)}{\phi^{(s)'}(0)} \tag{4.141}$$

and

$$Q = (1 - R)/(1 - R') \quad . \tag{4.142}$$

Our approach is to calculate Δ_0 in terms of the known static correlation functions (essentially an exact calculation) and then determine Q as an expansion in ε.

A direct calculation of $\phi^{(d)}$ and in turn R is rather difficult. One can show [4.16b] that $\phi^{(d)}$ (0,0) is nonpositive and, from stability considerations, $R \leq 1$. One can also show [4.6] that the high-temperature expansion for $\phi^{(d)}$ (0,0) starts at order A^4 (that is at fourth order in K). Finally one can show that R is zero to lowest order in our cell expansion in ε. We have then that

$$\Delta = \Delta_0[1 + O(\varepsilon)] \quad . \tag{4.143}$$

The quantity $\phi^{(s)}(q)\tilde{C}(q)$ can be calculated exactly,

$$\varphi^{(s)}(q)\tilde{C}(q) \equiv \alpha\Gamma^{(s)} = \alpha[1 + 4A\varepsilon(1,0) + 4A^2\varepsilon(1,1)] \quad , \tag{4.144}$$

where $\varepsilon(1,0)$ and $\varepsilon(1,1)$ are the known [4.18] nearest- and next-nearest-neighbor correlation functions. $\Gamma^{(s)}$ is independent of q and a rather rapidly decreasing positive-definite function of K. Near T_c, $\phi^{(s)}(q)$ is, therefore, dominated by the behavior of $\tilde{C}^{-1}(q)$. The characteristic frequency for the problem for small q and Z is $\omega_c = \phi(0,0)$, which should behave, near T_c, as $\omega_c \alpha \xi^{-z}$. Hence, when $\phi^{(d)}$ is neglected, we find $z = z_0 = (2 - \eta)$ (the "conventional" result). The bounds for

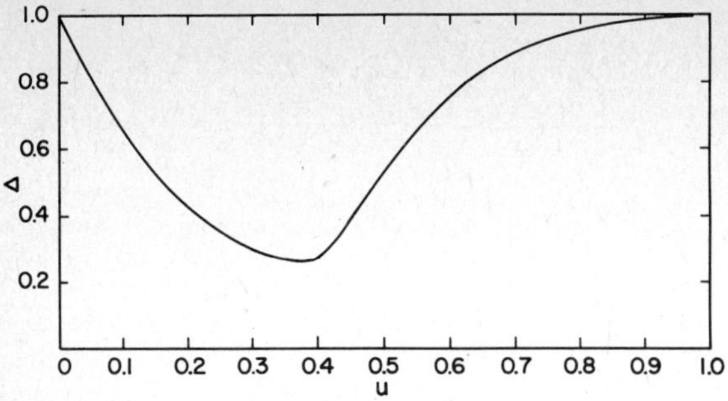

Fig.4.3. The ratio of characteristic times $\Delta = \alpha'/\alpha$ as given by (4.145), as a function of $u = \tanh K$. The minimum value occurs at a temperature slightly above T_c (note that $u_c = \sqrt{2} - 1$)

$\phi^{(d)}(0,0)$ and R discussed above imply that z_0 is a lower bound to z. Combining (4.141,143,144) we obtain our zeroth-order result for Δ_0,

$$\Delta \simeq \Delta_0 = \frac{X'}{X}\frac{\Gamma^{(s)}}{\Gamma^{(s)'}} \quad , \tag{4.145}$$

and, in turn,

$$\alpha_0 = \Delta_0\alpha/(1 - a_0) \quad . \tag{4.146}$$

We plot Δ_0 versus K in Fig.4.3. The dynamic index z is given in this case by inserting the value of Δ_0 at the critical point into (4.106). As expected we obtain the conventional value for $z = 2 - \eta$. Note that with this choice $\Delta_0 \to 1$ as $T \to 0$ and $\alpha_0 \to \infty$. This may look a bit peculiar. However, we note that the slowest mode $\lambda^{(1)} = \alpha_0(1 - a_0) = \Delta\alpha$ goes to α as $T \to 0$ and $\alpha_0 \to \infty$ which reflects the fact that all of the other modes for a cell, $\lambda^{(n)}$ for $n > 1$, decay extremely rapidly for low temperatures while the total spin $\psi^{(1)}$ decays rather slowly. We now have all the ingredients we need to solve the recursion relations. We present the results in the next section.

4.6 Results

We present here the results obtained by solving the recursion relations obtained in Sect.4.4 by direct iteration.

A rather simple check on the quality of our recursion relations involves the determination of the critical values of the short-range correlation functions $\varepsilon(2,0)$ and $\varepsilon(1,1)$ which satisfy the recursion relations (4.111b,c), respectively. We easily find that $\varepsilon(2,0) = 0.5988$ and $\varepsilon(1,1) = 0.6199$ at T_c. These compare favorably with the exact values $\varepsilon(2,0) = 1 - 4/\pi^2 = 0.5947$ and $\varepsilon(1,1) = 2/\pi = 0.6366$.

The next simplest quantity we can calculate is the magnetization. The most direct analysis follows from a straightforward iteration of (4.109). A more sophisticated analysis (see footnote 4) which includes the effects of spontaneous sym-

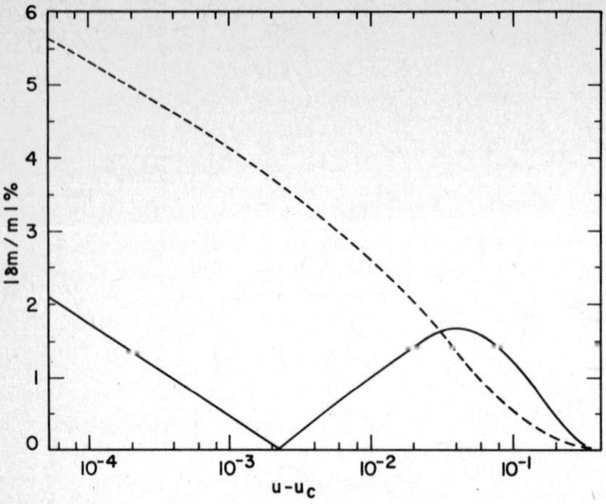

Fig.4.4. The difference between the calculated magnetization and the exact result m_E: $|\delta m/m| \equiv |(m-m_E)/m_E|$. The dashed line corresponds to the result of iterating (4.109), which is always larger than m_E. The solid line is the result of the analysis preformed by the authors, where m is larger than m_E very close to T_c, but smaller elsewhere (see footnote 4)

metry breaking obtains close agreement with the exact result [4.19] at all temperatures. In Fig.4.4 we display the percentage error in our results relative to the exact values of m. The symmetry breaking corrections alluded to above are incorporated in the result represented by the solid line, which is within $\sim 2\%$ of the exact result at all temperatures. The exponent β that we obtain is $\beta = 0.120$ ($\beta = 1/8$ is the exact result). The dashed line corresponds to (4.109).

The calculation of the magnetic susceptibility for $T > T_c$ follows from an iteration[5] of (4.125) and corresponds to the q = 0 values in Figs.4.5,6. As usual near T_c we can fit χ to a form $A\,t^{-\gamma}$ where $t = |K - K_c|/K_c$. We numerically obtain the exponent $\gamma = 1.76$ [compared to the exact result 1.75] both above and below T_c. We have extracted the amplitude A- for $T < T_c$ and find A- = 0.0249 (compared to the result 0.0255 obtained from series expansions [4.20]). We also plot $\tilde{C}(q)$ for $q_x = q_y = 0.1$ (in units of inverse lattice spacing) as a function of u in Figs.4.5,6. As q increases from zero to 0.1 we find that $\tilde{C}(q)$ has a maximum above T_c in agreement with [4.20] (a detailed comparison is given in [4.11]). The critical exponent η is found to be $\eta = 0.24$ ($\eta = 1/4$ is exact). The specific heat C [4.12] and the correlation length (see footnote 4) can be found in a similar way. C has a finite cusp at T_c, rather than a divergence. The correlation length diverges at T_c with an exponent $\nu = 0.99998$ ($\nu = 1$ exact). The amplitude associated with this divergence below T_c is $A_\xi^- = 0.1485$. The result from series expansions [4.20] is $A_\xi^- = 0.175$.

We now turn to the dynamics. It follows from our scaling analysis in Sect.4.4 that the dynamic critical exponent z is $z = \gamma = 1.76$ to this order. The reliability of this result is treated in detail in [4.6]. The dynamic correlation func-

5 Equation (4.125) can be rewritten in the form $\chi - 1 = 4\nu_1^2 - 1 + 4\nu_1^2(\chi' - 1)$. When we iterate this equation (solve for $\chi-1$), we find that the sequence converges rapidly for $T > T_c$ since as K scales to 0 χ scales to 1.

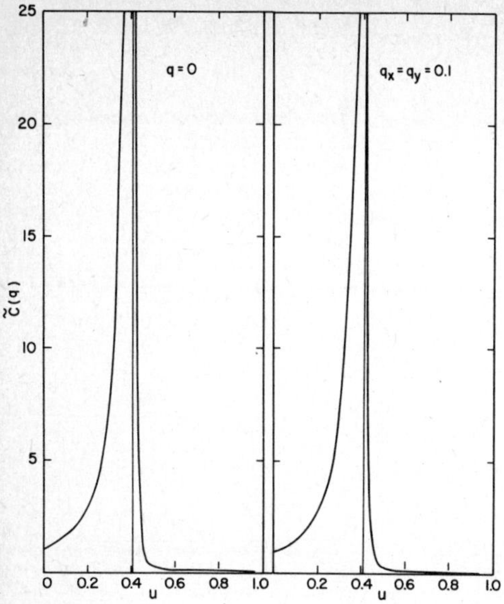

Fig.4.5. The static structure factor $\tilde{C}(q)$ obtained from (4.124) plotted as a function of $u = \tanh K$ over the entire temperature range for two values of q. Note that $\tilde{C}(q=0)$ is the susceptibility χ. The vertical line marks the position of $u_c = \sqrt{2} - 1$

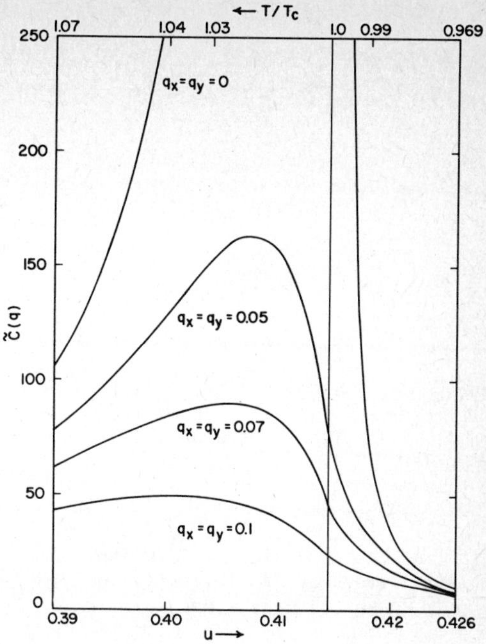

Fig.4.6. The static structure factor $\tilde{C}(q)$ (4.124) for four values of q, at temperatures near T_c (notice the upper horizontal scale). $\tilde{C}(q)$ passes smoothly through T_c (vertical line) and, for $q \neq 0$, has a maximum at some $T > T_c$

tions can, however, be evaluated at all temperatures using the recursion relation (4.117). It is instructive to examine first the same-site correlation function $C_{ii}(t)$. Its recursion relation is a particular case of (4.115). In Fig.4.7 we have the results for $T > T_c$. Close to T_c, $C_{ii}(t)$ decays to zero extremely slowly. As the temperature increases, however, $C_{ii}(t)$ decays progressively faster. As $K \to 0$ it decays exponentially $[C_{ii}(t) = \exp(-\alpha t)]$ as indicated by the dashed line. Below T_c, we obtain the results shown in Fig.4.8 where we have plotted $C_{ii}(t)/C_{ii}(0)$ (note that $C_{ii}(0) = 1-m^2$) at several temperatures. Again we observe the critical slowing down as we approach the critical point.

We turn next to the dynamic structure factor $C(q,\omega)$ which we plot in Fig.4.9 as a function of ω and fixed q for several temperatures above T_c. A plot of $C(q,\omega)$ at finite q, but for temperatures below T_c is shown in Fig.4.10. These figures illustrate the variation of $C(q,\omega)$ with ω and K. The variation with q is illustrated in Fig.4.11, where the temperature is held constant and $C(q,\omega)$ is plotted for several values of q.

The behavior near T_c is shown in Fig.4.12. The straight line is $C(0,\omega)$ at T_c which diverges as $\omega^{-(2+z-\eta)/z}$ (ω^{-2} since $z = 2 - \eta$ in our case) in agreement with

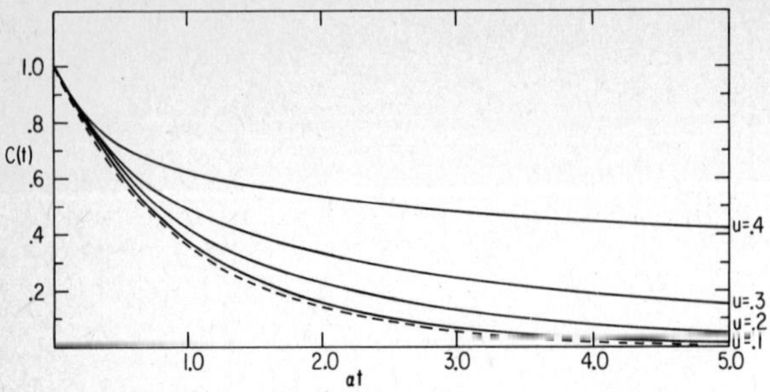

▲ Fig.4.7. The same-site correlation function $C_{ii}(t)$ versus time at several $T > T_c$. The dashed line is the high-temperature limit

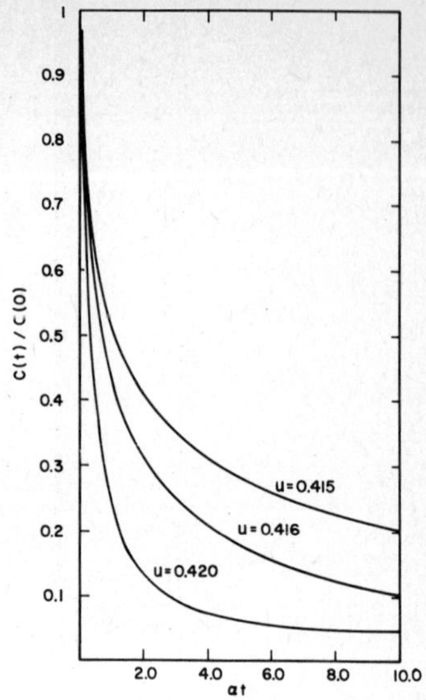

Fig.4.8. The same-site correlation function $C_{ii}(t)$ divided by its initial value $C_{ii}(0) = 1-m^2$ versus time at several $T < T_c$ ►

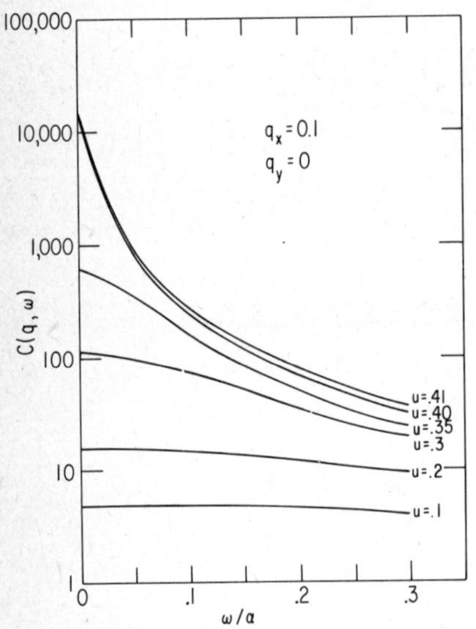

Fig.4.9. The dynamic structure factor $C(q,\omega)$, obtained from (4.116,122), at fixed q for several $T > T_c$ as a function of frequency ω

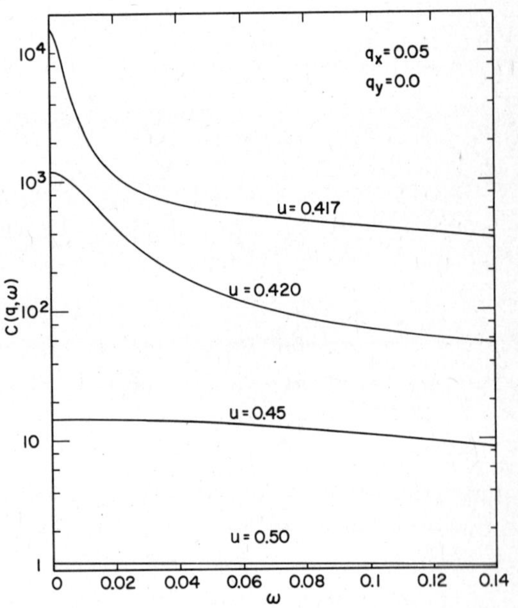

Fig.4.10. $C(q,\omega)$ at finite q as in Fig.4.9 but at several $T < T_c$

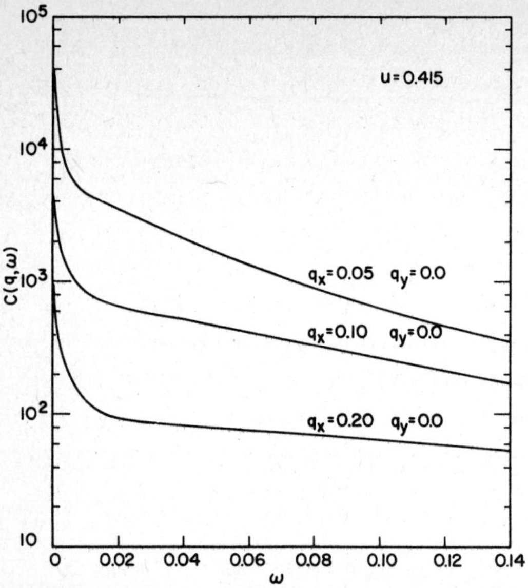

Fig.4.11. $C(q,\omega)$ as a function of ω and at constant temperature for several values of q

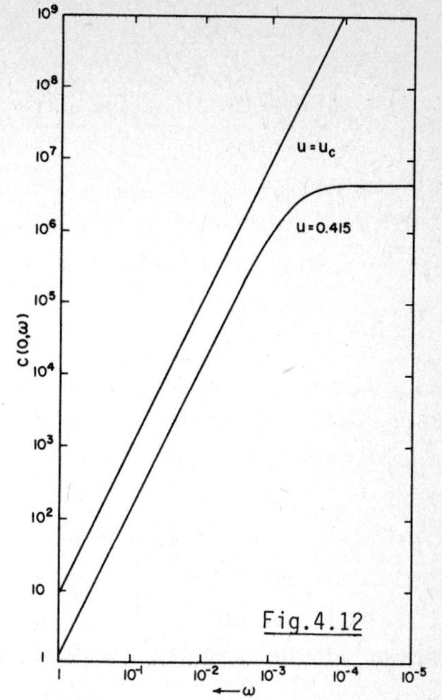

Fig.4.12. The dynamic structure factor near T_c and $q = 0$. The straight line is the $T = T_c$ limit and its slope is 2, in a agreement with dynamic scaling (see text). At $T \gtrsim T_c$ ($u = \tanh K = 0.415$) $C(0,\omega)$ is a Lorentzian

scaling. Near T_c, $C(0,\omega)$ is a Lorentzian (see [4.12] for details).

As we have discussed in previous work [4.11], all of the expected scaling relations in the asymptotic critical regime are satisfied. It is also possible to evaluate the scaling functions which turn out to be of Lorentzian shape in both the scaling and the hydrodynamic regimes.

4.7 Discussion

The results of the last section indicate that our RSDRG methods can be used to evaluate space- and time-dependent correlation functions over a wide range of temperatures, wave numbers and frequencies. The results obtained compare very well with known exact or series results.

The model we have studied is rather simple. As we have pointed out the conventional theory which corresponds to neglecting $\phi^{(d)}$ in (4.135) should give a rather good description of the dynamics for $T > T_c$ except, possibly, for a very narrow dynamic critical region. For $T > T_c$, one can take an alternative point of view to that expressed in the body of this chapter and reflected in the iterative solution of (4.117). This alternate point of view says that our RG description is appealing mainly because it provides an extremely convenient method for obtaining the needed static results that go into $\phi^{(s)}(q)$ [see (4.144)] and $C(q,z)$ then follows from (4.135) with $\phi^{(d)} = 0$. We have found, for the most part, that these points of view are in close agreement.

The situation for $T < T_c$ is much less clear. We are not aware of any low-temperature expansions treating the dynamics. The development here is considerably more involved than in the carrying out high-temperature expansions. We have found, however, that our procedures seem to work equally well in the ordered or disordered phases.

We believe that the extension of these ideas to more complicated models (such as those with a conserved magnetization) will be very useful and informative since other methods are not available for treating these problems. In particular in models with spin exchange it is rather difficult to carry out a high-temperature expansion beyond the lowest order since even at infinite temperature there is a dynamic coupling between spins due to the exchange process. There are no such difficulties associated with our block spin methods.

Finally we point out our belief that these ideas can be extended to treat fully nonequilibrium problems (such as deep temperature quenches). This development would open the way to the exciting possibility of analytical studies of the kinetics of first-order phase transitions.

Acknowledgements. GFM acknowledges support from an Alfred P. Sloan Foundation Fellowship and from the National Science Foundation (NSF DMR 80-20609). OTV was supported by the University of Minnesota Graduate School Research Fund.

References

4.1 S. Ma: *Modern Theory of Critical Phenomena* (Benjamin, Reading 1976); D. Amit: *Field Theory, The Renormalization Group and Critical Phenomena* (McGraw-Hill, New York 1978); G. Toulouse, P. Pfeuty: *Introduction au groupe de renormalisation et a ses applications* (Presse Universitaire, Grenoble 1975)

4.2 K.G. Wilson: Phys. Rev. B*4*, 3174, 3184 (1971)

4.3 Th. Niemeijer, J.M.J. van Leeuwen: In *Phase Transitions and Critical Phenomena*, Vol.6, ed. by C. Domb, M.S. Green (Academic, New York 1976)

4.4a M. Nauenberg, B. Nienhuis: Phys. Rev. Lett. *33*, 1598 (1974);

4.4b L. Kadanoff, A. Houghton, M.L. Yalabik: J. Stat. Phys. *14*, 171 (1976)

4.5 Y. Achiam, J. Kosterlitz: Phys. Rev. Lett. *41*, 128 (1978); Y. Achiam: J. Phys. A*11*, L129 (1978); A*13*, 1355 (1980); W. Kinzel: Z. Phys. B*29*, 361 (1978); M. Suzuki, K. Sogo, I. Matsuba, H. Ikada, T. Chikama, H. Takano: Prog. Theor. Phys. *61*, 864 (1979); S.T. Chiu, G. Forgacs, H.L. Frisch: Phys. Rev. B*20*, 243 (1979); U. Dekker, F. Haake: Z. Phys. B*36*, 379 (1980); S. Ma: Phys. Rev. B*19*, 4824 (1979)

4.6 G.F. Mazenko, O.T. Valls: Phys. Rev. B*24*, 1419 (1981)

4.7 G.F. Mazenko, M.J. Nolan, O.T. Valls: Phys. Rev. Lett. *41*, 500 (1978)

4.8 G.F. Mazenko, M.J. Nolan, O.T. Valls: Phys. Rev. B*22*, 1263 (1980)

4.9 G.F. Mazenko, M.J. Nolan, O.T. Valls: Phys. Rev. B*22*, 1275 (1980)

4.10 G.F. Mazenko, J.E. Hirsch, M.J. Nolan, O.T. Valls: Phys. Rev. Lett. *44*, 1083 (1980)
4.11 G.F. Mazenko, J.E. Hirsch, M.J. Nolan, O.T. Valls: Phys. Rev. B*23*, 1431 (1981)
4.12 G.F. Mazenko, O.T. Valls: Phys. Rev. B*24*, 1404 (1981)
4.13 G.F. Mazenko, E. Oguz: Journ. Stat. Phys. (in press)
4.14 G.F. Mazenko, J. Luscombe: Ann. Phys. N.Y. *132*, 121 (1981)
4.15 G.F. Mazenko: Unpublished
4.16a R.J. Glauber: J. Math. Phys. N.Y. *4*, 294 (1963);
4.16b K. Kawasaki: In *Phase Transitions and Critical Phenomena*, Vol.2, ed. by C. Domb, M.S. Green (Academic, New York 1972)
4.17 J. Luscombe, G.F. Mazenko: Unpublished
4.18 B. Kaufmann, L. Onsager: Phys. Rev. *76*, 1244 (1949)
4.19 C.N. Yang: Phys. Rev. *85*, 808 (1952)
4.20 H. Tarko, M. Fisher: Phys. Rev. B*11*, 1217 (1975)

5. Renormalization for Quantum Systems

P. Pfeuty, R. Jullien, and K. A. Penson

The quantum nature of physical systems shows up at very low temperatures and affects phase transition phenomena. The existence of quantum transitions at T = 0 leads to the manifestation of quantum-classical crossover phenomena in classical transitions at low temperatures. The real space renormalization group methods introduced for classical systems have been extended to quantum systems at $T \neq 0$. At T = 0 a block renormalization group method has been introduced to study the ground state and the excited states of a many-body quantum system and to analyze transitions which take place in the ground state of the system. These different methods are described and discussed, giving more emphasis to the block method. The numerous applications to spin and fermion systems and to field theory in one and higher dimensions are presented.

5.1 Background

Phase transitions have been studied mainly in classical systems [5.1-3]. The Ising model has been solved exactly in two dimensions [5.4], and the isotropic n-vector model has been solved in one dimension [5.5]. Landau-Ginzburg free-energy functionals have been studied using renormalization group diagrammatic techniques [5.3].

The *quantum nature* of physical systems cannot be ignored *at very low temperature*. Universality ideas which have emerged from early studies of phase transitions [5.2] lead to the fact that the critical behavior should be affected by quantum effects if the transition takes place at low temperature.

The simplest quantum model which presents a second-order phase transition is the Ising model with a transverse field (ITF) [5.6,7]. This model with the O(1) or Z(2) symmetry is a nontrivial extension of the Ising model [5.4] which has been introduced as a simple model for some ferroelectric and magnetic systems [5.6,7]. It was first shown [5.8] that the critical behavior, if the transition temperature is finite, should not be affected by the quantum effects due to the transverse field and should be the same as for the classical Ising model. This property was checked later [5.9] from high-temperature series expansions study. At T = 0 a transition takes place for a critical value Γ_c of the transverse field [5.10]. This "quantum" transition corresponds to a drastic change of the ground state which becomes asymptotically degenerate for $\Gamma < \Gamma_c$. The critical behavior at T = 0 when $\Gamma \to \Gamma_c$ is totally different from the classical behavior at $T \neq 0$. For instance, in one dimension there is no transition for the classical Ising model and the quantum ITF model at T = 0 shows a transition which can be studied exactly [5.11]. At low temperature there are quantum-classical crossover phenomena [5.12]: when T is fixed and Γ approaches $\Gamma_c(T)$, the critical behavior changes from a quantum to the asymptotic classical form [5.13].

These ideas have been generalized by HERTZ [5.14]. Any quantum problem can be formulated with a Landau-Ginzburg functional integral where the time variables (or frequencies) are included and the usual renormalization group diagrammatic techniques can be used. This approach was applied [5.15] to the ITF model and later [5.16,17] to the quantum-classical crossover.

At the same time, the real-space renormalization group method introduced [5.18] for Ising-like classical systems was extended to quantum systems such as the spin 1/2 XY model at finite temperature [5.19-25], the Ising model in a transverse field at T = 0 [5.26-28] and at $T \neq 0$ [5.29-30]. In this method the lattice is divided into cells. The Hamiltonian is separated into an intracell and an intercell part treated in perturbation; partial traces are made and new "spins" are associated with each cell using a majority rule [5.18,19] or some other more general rule [5.23-25]. The Migdal-Kadanoff method was also applied to quantum systems [5.31]. It was not clearly realized that the conventional real-space renormalization group methods developed for classical systems were not well suited for quantum systems in the low-temperature quantum regime. At low and especially at zero temperature another method has to be used, which allows building the ground state and the low-energy excitations adequately. Such a method was first applied to the Kondo model of magnetic impurities in metals [5.32]. An iterative scheme is used where, at each step, the highest energy levels are neglected and the lowest retained to build the ground state and the lowest excited states. The only other work in the same line has been the extension to the Anderson model [5.33]. In 1976, two groups – one at SLAC (field theory) and one in Holland (solid-state theory) – rediscovered the same method.

At SLAC in Stanford, a block renormalization method was proposed [Ref.5.34 , especially Sect.IV.D] to study the ground state for the Hamiltonian formulation of field theory on a lattice. They wanted to develop a nonperturbative technique to understand the quark confinement which implies solving a nonabelian gauge theory of interacting fermion quarks and vector mesons. This method was first applied as a test to two simple models [5.35]: the Ising model in a transverse field and the free spinless field theory in one-time one-space dimensions. The SLAC group applied the same method to QED field theory [5.36], fermion models [5.37], and finally QCD field theory [5.38]. It was extended by others to study various lattice field theory models (or statistical mechanics lattice models) such as gauge models [5.39-41],

Potts P(n) models [5.42-44], Ising-like models [5.40,45-63], and solid-state quantum problems including spin [5.64-79] and fermion [5.80-88] systems. This method is used to calculate ground-state energies and to study quantum critical phenomena at transitions in the ground state.

Independently a Dutch group introduced the same method but focused mainly on the determination of the ground-state energy for low-dimensional spin 1/2 Heisenberg systems [5.89] as presented in a recent review [5.90].

Quantum phase transitions are of interest for two reasons:

a) when a system presents a transition at low temperature quantum fluctuations may lead to new kinds of phase transitions,

b) there is a mathematical correspondence [5.91] between statistical mechanics and quantum mechanics where time plays the role of an extra dimension. This was clearly shown for Ising systems [5.11,92-96] and generalized [5.97,98]: any statistical mechanics system on a lattice in d dimensions can be considered as isomorphic (it has the same transition) to a quantum-mechanical Hamiltonian system on a lattice in d-1 dimensions, the extra dimension being replaced by $i \cdot t$ where t is the time. This idea has been much used by field theorists recently to study the lattice field theory.

In this review, we shall only focus on the application of real-space renormalization group methods to quantum systems. Other methods such as exact solutions, series expansions, perturbative approaches are not considered here. In Sect.5.2 we present the extension of the N-VL [5.18] renormalization group method to quantum systems mainly at T = 0. In Sect.5.3 the block method specific to quantum systems is presented. Section 5.4 is concerned with the numerous applications of the block method. Section 5.5 contains a discussion. In this last section the closely related finite-size scaling method recently applied to quantum systems [5.99-5.105] is compared to the RG method.

5.2 Application of the Niemeijer-van Leeuwen Renormalization Group Method to Quantum Lattice Models

Let us first review very briefly the Niemeijer-Van Leeuwen renormalization group method [5.18]. This method has been applied to discrete spin systems and mainly to two-dimensional spin 1/2 Ising lattice systems. It concerns the transformation of a discrete spin system to another discrete spin system on an isomorphic lattice. Let us consider the standard example: an Ising system on a triangular lattice. Triangular cells are drawn so that the cell system forms again a triangular lattice. A new cell spin with value ±1 is associated with each cell (the value of the cell spin depends on the value of the site spins of the cell following a rule;

the majority rule is that for an odd number of spins in the cell the cell spin is equal to the sign of the sum of the site spins of the cell). The quantity of interest in equilibrium statistical mechanics is the partition function Z of the original spin system which is first computed with the constraint of a given configuration of cell spins. The resulting free energy depends on the cell-spin configuration and can be seen as deduced from an effective cell-spin Hamiltonian H'. A transformation is achieved from the site-spin Hamiltonian H to a cell-spin Hamiltonian H' defined on a lattice where the unit length is larger so that the length has been reduced by a factor ρ ($L' = L/\rho$). This defines a renormalization group transformation, and the general formalism of scaling and renormalization can be used to deduce the critical point properties of the system. This strategy cannot be done exactly (up to now) as far as we know, and even for the two-dimensional Ising model only approximate techniques have been developed.

The eight configurations of the site spins S_1, S_2, S_3, of a cell can be associated with a cell spin $S_i' = \pm 1$ and an index i = 1, 2, 3, 4. The value of S_i' is equal to one times the sign of the sum $(S_1 + S_2 + S_3)$. The new Hamiltonian H' is defined by

$$\exp[H'(S')] = \sum_{\sigma} \exp[H(S';\sigma)] \tag{5.1}$$

where $\beta = 1/kT$ has been incorporated in H. Equation (5.1) cannot be solved exactly. Perturbation methods can be used to solve it. The Hamiltonian H is divided into two pieces

$$H = H_0 + V \quad , \tag{5.2}$$

where H_0 is an intracell part and V an intercell part. Equation (5.1) can be rewritten

$$\exp[H'(S')] = \sum_{\sigma} \exp[H_0(S';\sigma)] \quad \langle \exp V \rangle_0 \tag{5.3}$$

where

$$\langle \exp V \rangle_0 = \frac{\sum_{\sigma} \exp[H_0(S',\sigma)]\exp(V)}{\sum_{\sigma} \exp H_0(S';\sigma)} \quad . \tag{5.4}$$

A cumulant expansion is used for $\langle \exp V \rangle_0$,

$$\langle \exp V \rangle_0 = \langle V \rangle + \frac{1}{2} (\langle V^2 \rangle - \langle V \rangle^2) + \ldots . \tag{5.5}$$

To first order in V the calculation is easy and

$$H'(S') = N \ln[\exp(3K)+3 \exp(-K)]+2K \left(\frac{\exp(3K)+\exp(-K)}{\exp(3K)+3\exp(-K)}\right)^2 \sum_{ijnn} S_i'S_j' \tag{5.6}$$

with $H(S) = -K \sum_{ijnn} S_iS_j$, H' has the same form as H with only nearest-neighbor interactions. To next order in V new interactions appear so that the problem becomes more and more complicated.

This simple approximate method has been extended to spin 1/2 quantum models defined on a two-dimensional lattice: the spin 1/2 XY and Heisenberg model [5.19] and the anisotropic spin 1/2 Heisenberg model [5.22]. The Hamiltonian is divided into two pieces H_0 and V as before, but (5.4) is replaced by

$$\exp(H_0 + V) = \exp(H_0)\exp(W)$$

with

$$W = V + \frac{1}{2}[V,H_0] - \frac{1}{12}[2H_0 + V,(V,H_0)] + \dots \quad . \tag{5.7}$$

The cumulant expansion (5.5) is then used with V replaced by W. This method has been mainly applied to the quantum spin 1/2 XY model on the triangular lattice with cells of three spins [5.19] and with cells of seven spins [5.21], and for the quadratic lattice with cells of five spins [5.20]. In no case could a nontrivial fixed point of the renormalization transformation be found in the first order of a cumulant expansion, while a fixed point appeared in the second order only for the triangular lattice but with unphysical critical exponents ($\delta < 0$). This model has also been applied to the ITF model on a triangular lattice. The Ising-like critical behavior is recovered at nonzero temperatures [5.29].

Other renormalization schemes have been proposed to get better results. For instance, the majority rule has been extended to include the three spin component operators S^x, S^y, S^z [5.23]. Equation (5.1) is rewritten

$$\exp[H'(S')] = \mathop{\mathrm{Tr}}_{S}\{\exp H(S)P(S',S)\} \tag{5.8}$$

with

$$P(S',S) = \mathop{\pi}_{\alpha}\left[\frac{1}{2}\left(1+S'^{x}_{\alpha}\,\mathrm{sign}\left\{\sum_{\ell} S^{x}_{\ell,\alpha}\right\}+S'^{y}_{\alpha}\,\mathrm{sign}\sum_{\ell} S^{y}_{\ell,\alpha}+S'^{z}_{\alpha}\mathrm{sign}\sum_{\ell} S^{z}_{\ell,\alpha}\right)\right] \quad . \tag{5.9}$$

Another general transformation has been proposed [5.25] and applied to the anisotropic Heisenberg model on a quadratic lattice. The weight operator P(S',S) is supposed to have the form

$$P(S',S) = \mathop{\pi}_{\alpha}\left\{\frac{1}{2} + \underline{S}'_{\alpha}\cdot\left(\sum_{\ell}\underline{S}_{\alpha,\ell}\right)Q\left[\left(\sum_{\ell}\underline{S}_{\alpha,\ell}\right)^{2}\right]\right\} \quad . \tag{5.10}$$

Hoping to get better results, a real-space linear renormalization group transformation has been proposed for studying the critical properties of the two-dimensional spin 1/2 XY model on the triangular lattice [5.24]. In this transformation, the weight operator P(S',S) is linear and takes the form

$$P(S',S) = \mathop{\pi}_{\alpha}\left[\frac{1}{2} + pS'^{z}_{\alpha}\left(\sum_{\ell} S^{2}_{\alpha,\ell}\right)\right] \quad . \tag{5.11}$$

This transformation depends on the parameter p which is fixed so that marginality for temperature deviations from criticality is obtained as predicted for the classical XY model and the magnetic exponent η is found close to the expected value ($\eta = 1/4$).

The method presented above can also be applied at T = 0 in the following way [5.26]. At zero temperature T = 0, the thermodynamic properties are replaced by the ground-state properties. Let us consider again, for example, a spin 1/2 quantum system defined on a triangular lattice. The latter is divided into triangular cells of three sites. The new lattice of cells is triangular and contains N/3 sites (if the original lattice has N sites). The ground-state properties are contained in the projection operator $P(H) = |0><0|$ where $|0>$ is the ground state. It is enough to know the diagonal elements of P(H) on a simple basis. We can choose as a basis state the product $|S_1^z>|S_2^z>\dots|S_N^z>$ where $|S_i^z>$ is an eigenstate of the operator S_i^z. This state can also be written as a product of N/3 cell states $|\alpha>$ where each cell state is a product of three site states. The 2^3 cell states are divided into two groups. A cell state is written $|S'^z_\alpha, q_\alpha>$ ($S'^z = \pm 1$, $q = 1,\dots 4$), and a majority rule is used again so that $S'^z_\alpha = \text{sign} \sum_\ell S^z_{\alpha,\ell}$. The renormalization equation (5.1) now becomes

$$<S'^z_1|\dots<S'^z_{N/3}|0>'<0'|S'^z_{N/3}\dots|S'^z_1>$$

$$= \sum_{q_\alpha} <S'^z_{1,q_1}|\dots S'^z_{N/3,q_{N/3}}|0><0|S'^z_{N/3,q_{N/3}}\dots|S'^z_{1,q_1} \quad . \tag{5.12}$$

The new ground state $|0>'$ defined by (5.12) is the ground state of a new Hamiltonian H' defined on a lattice of N/3 sites where the lattice spacing has been increased by a factor $\sqrt{3}$. Equation (5.12), which leads to the renormalization transformation

$$H \to H' \quad ,$$

cannot be solved exactly, a perturbation scheme is used to solve it. The Hamiltonian H is separated into an intracell part H_0 and an intercell part V which is treated in perturbation. To first order the projection operator P(H) is equal to

$$P(H) = |0>_{00}<0| + \sum_{n\neq 0} \frac{|0>_{00}<0|V|n>_{00}<n|}{E_0 - E_n} + \sum_{n\neq 0} \frac{|n>_{00}<n|V|0>_{00}<0|}{E_0 - E_n} \tag{5.13}$$

where $|n>_0$ and E_n are the eigenstates and eigenvalues of H_0. The projection operator P(H') is also expanded in perturbation so that H' is obtained from (5.12,13). In general, H' is different from H so that the parameter space is enlarged. This method has been applied mainly to the ITF model both in one dimension [5.27,28] and in two dimensions [5.26,27]. To first order H' takes the same form as H so that only one recursion relation is obtained. To second order next-nearest-neighbor and next-

next-nearest-neighbor interactions are also generated. In the first study [5.26], the critical transverse field $(\Gamma/J)_c$ and the critical exponent ν were determined. Later [5.27] both the ground-state energy in a longitudinal field and the two exponents ν and δ were calculated. At $T = 0$ the dynamic exponent z which tells how energy is rescaled also has to be determined from the low-energy excitations. This can be done by applying the same method at finite temperatures as at low temperatures (in [5.27] this was not done because of the computational difficulties; instead, z was taken equal to the expected value $z = 1$ [5.92,98]). The results obtained for the critical properties of the two-dimensional ITF model can be compared with the series expansion results [5.10] (ν is found to be equal to 0.967 to first order and 0.924 to second order as $\nu = 0.63$ from series expansions; δ is found to be equal to 1.13 to first order and 4.4 to second order as $\delta \simeq 5$ from series expansions). In one dimension [5.28] the results obtained with cells of three spins give rather bad results especially for the exponent δ(ν is found to be equal to 1.13 to first order and 0.95 to second order to be compared with $\nu = 1$ from the exact solution [5.11]; the exponent δ is found to be equal to 2.18 to first order and -9.6 to second order when the exact solution gives $\delta = 15$). No systematic study of the dependence on the cell size has been made. These preliminary results on the ITF model are not very good. A similar calculation for the ITF model on a quadratic lattice using four spin cells both at $T = 0$ and at $T \neq 0$ gives $\nu = 0.72$ at $T = 0$ and allows studying the classical quantum crossover [5.30].

This method suffers from the difficulty of choosing the renormalization transformation which defines the cell operators out of the spin operators. Another method has been developed especially for $T = 0$ where this choice is dictated by the rule that the ground-state energy has to be the lowest energy. We shall try to develop this method in the next section.

5.3 The Block Method

5.3.1 Principles

Let us consider a Hamiltonian H defined on a lattice L with N sites. The first step is to divide the lattice L into N/n_s blocks of n_s sites each, so that these cells form a new lattice L' which has the same geometry as L (for instance, if L is a triangular lattice of spacing a then L' is also a triangular lattice of spacing $a' = a \times \sqrt{n_s}$). The Hamiltonian H can then be divided into two pieces H^0 and V where H^0 is the block Hamiltonian and V contains the coupling part between different blocks.

The second main step is to diagonalize H^0 exactly, using both analytic symmetry properties and numerical analysis. The eigenstates of H^0 are then separated into two classes: the i states and the remaining α states. The i states, which are usually the lowest ones, are retained to form a new basis for a new Hamiltonian H' defined on the new lattice L'. We can write

$$H^0 = \sum_i |i> \varepsilon_i < i | + \sum_\alpha |\alpha> \varepsilon_\alpha <\alpha| \quad . \tag{5.14}$$

The new Hamiltonian $H' = H'^0 + V'$ is such that the "i" eigenvalues of H and H' are the same. The basic equations for the renormalization group recursion relations $H \to H'$ are

$$H'||\mu_i>> = E_i||\mu_i>> \quad ,$$

$$H ||i>> = E_i || i>> \quad , \tag{5.15}$$

where the states $||i\gg$ are the eigenstates of H which go to $|i>$ when V goes to zero, and where the states $||\mu_i\gg$ are the eigenstates of H' which go to $|\mu_i>$ when V' goes to zero with $H'^0 = \sum_i |\mu_i> \varepsilon_i <\mu_i|$. In most cases (5.15) cannot be solved exactly so that perturbation schemes have to be used. V and V' are treated in perturbation in (5.15). Depending on the choice of the n_L retained levels in each block and of the order in perturbation used, the new Hamiltonian H' generally has a more complicated form than H so that the parameter space is enlarged and increases rapidly when the order of perturbation increases. Recursion relations for coupling constants are obtained and analyzed to give information about the ground-state properties (energy, gap, etc.) and the quantum-phase transitions at T = 0 for critical values of the coupling constants. Following the well-known renormalization group formalism [5.3] the critical exponents can be extracted. The dynamical exponent z related to the rescaling of time or energy when length is rescaled is also obtained (the time dependence is governed by the quantum nature of the problem). To show how the method works we shall apply it to two simple "toy" models in one dimension: the ITF model and the free fermion model. Both first- and second-order perturbation treatments are given.

5.3.2 Applications

a) The Ising Model in a Transverse Field in One Dimension

In one dimension the Ising model in a transverse field is represented by the Hamiltonian

$$H = - J \sum_i S_i^x S_{i+1}^x - h \sum_i S_i^z \quad . \tag{5.16}$$

where the operators S_i^x and S_i^z defined on each site i of a chain of N sites (N large) are spin 1/2 operators which can be represented by Pauli matrices

$$S_i^z = \begin{pmatrix} 1 & 0 \\ 0 & -1 \end{pmatrix} \quad , \quad S_i^x = \begin{pmatrix} 0 & 1 \\ 1 & 0 \end{pmatrix} \quad . \tag{5.17}$$

We shall follow the renormalization block method presented in Sect.5.3.1. The chain of N sites is divided into N/2 blocks of $n_s = 2$ sites. H is divided into the intra-block part H^0 and the interblock part $V(H = H^0 + V)$ with

$$H^0 = \sum_j H_j \quad , \qquad V = \sum_j H_{j,j+1} \quad ,$$

with

$$H_j = - JS^x_{j,1}S^x_{j,2} - h(S^z_{j,1} + S^z_{j,2}) \tag{5.18}$$

and

$$H_{j,j+1} = - JS^x_{j,2}S^x_{j+1,1} \quad .$$

In the notation (j,p), j is the block index and p is the site index inside the block (p = 1, 2 here). H^0 is diagonalized exactly. For a block there are four eigenstates,

$$|0> = \frac{1}{(1 + a^2)} \{|++ > + a\,| -- >\} \quad ,$$

$$|1> = \frac{1}{\sqrt{2}} \{|+- > + | -+ > \quad ,$$

$$|2> = \frac{1}{\sqrt{2}} \{|+- > - | -+ > \quad ,$$

$$|3> = \frac{1}{(1 + a^2)} \{a\,| ++ > - | -- >\} \quad , \tag{5.19}$$

with the respective energies E_0, E_1, $- E_1$, $- E_0$ where

$$\begin{cases} E_0 = - \sqrt{4h^2 + J^2} \\ E_1 = - J & (5.20) \\ a = (\sqrt{4h^2 + J^2} - 2h)/J \quad . & (5.21) \end{cases}$$

The next step is to retain from these four states only the two lowest, |0> and |1>, so that the states |i> are of the form |0>, |0>, |1> ... |0>, and the |α> states of the form |2>, |2>, |3>, |2> ... . The new basis states $|\mu_i>$ are thus of the form |+>, |+> ... |->, |+> ... . In that basis the Hamiltonian $H'^0 = \sum|\mu_i>\varepsilon_i<\mu_i|$ takes the form

$$H'^0 = \sum_j (- h'S'^z_j + E') \tag{5.22}$$

with

$$\left.\begin{aligned} h' &= \frac{E_1 - E_0}{2} = \frac{\sqrt{4h^2 + J^2} - J}{2} \\ E' &= \frac{E_1 + E_0}{2} = -\frac{\sqrt{4h^2 + J^2} + J}{2} \end{aligned}\right\} \text{and} \quad . \tag{5.23}$$

The next step is to obtain H' from (5.16). To zeroth order $H' = H'^0$. To first order in perturbation V' takes the form

$$V' = -J' \sum_j S'^x_j S'^x_{j+1} \tag{5.24}$$

where

$$J' = J(1+a)^2/2(1+a^2)$$

and where S'^x_j is a Pauli matrix defined in the space μ_i. The new Hamiltonian $H' = H'^0 + V'$ has the same form as H with only a new constant term. The recursion relations are given by (5.23,24). This scheme is iterated many times until a fixed point is reached. The parameters take successive values $h^{(1)}, h^{(2)} \dots h^{(n)}, J^{(1)} \dots J^{(2)} \dots J^{(n)}, h^{(0)} = h, J^{(0)} = J$. There are two trivial stable fixed points $h = 0$ and $J = 0$ separated by a nontrivial unstable fixed point $(h/J) = (h/J)_c = 1.2767$. When $h/J < (h/J)_c$, $(h/J)^{(n)}$ tends to zero, and the ground state is a doublet. When $h/J > (h/J)_c$, $(h/J)^{(n)}$ tends to ∞; the ground state is then a singlet.

This method allows two types of calculation. a) It allows calculating physical quantities for all values of the parameters. b) It allows determining critical exponents when applied in the vicinity of the fixed point.

a) The quantities which can be calculated are the ground-state energy, the gap, the ground-state average $\langle S^\alpha_i \rangle$, and the correlation functions $\langle S^\alpha_i S^\alpha_j \rangle$. We shall here only consider the ground-state energy and the "magnetization" $\langle S^x_i \rangle = M^x$.

The ground-state energy. By iterating (5.23,24) several times we get the following expression for the ground-state energy per site E_G,

$$E_G = -\left(\sum_{n=0}^{\infty} \frac{J^{(n)} + \sqrt{4h^{(n)^2} + J^{(n)^2}}}{2^{n+1}} \right) \frac{1}{2} \ . \tag{5.25}$$

We find an energy larger than the exact one [5.11] by 30%. The second derivative of E_G with respect to h, $\partial^2 E_G/\partial h^2$, is shown in [Ref.5.45, Fig.2] and shows a singular behavior at $(h/J)_c$.

The magnetization, $\langle S^x \rangle$. Knowing the correspondence between the basis states of H' and of H it is easy to determine the relation between operator expectation values on L and on L'. With

$$M^{x(n)}_i = \langle S^{x(n)}_i \rangle$$

$$= \langle S^{x(n)}_{j,P(n)} \rangle \quad \text{and} \quad M^{x(n+1)}_j = \langle S^{x(n+1)}_j \rangle$$

we have

$$M^{x(n)}_i = \xi^{(n)}_{p(n)} \quad M^{x(n+1)}_j \quad .$$

After many iterations $M^{x(0)}_i = \xi^0_{p(0)} \xi^1_{p(1)} \cdots \xi^n_{p(n)} \ M^{x(n+1)}$, where in our case

$$\xi_1^{(n)} = \xi_2^{(n)} = \frac{1 + a}{\sqrt{2}(\sqrt{1 + a^2})} \quad . \tag{5.26}$$

For very large n, $M^{x(n)}$ is either 1 or zero (1 if the fixed point is at h = 0 and 0 if the fixed point is at J = 0). By using (5.26) M_i^x can be extracted; it depends on its location i but for $n_s = 2$ it is unimportant. For larger n_s, site i has to be chosen so that the edge effect of the block partitioning can be minimized. The magnetization curve obtained by this method is shown in [Ref.5.45, Fig.3].

Critical exponents. Close to the critical point $h/J \sim (h/J)_c$ the correlation length of the system diverges,

$$1 \sim \left[\frac{h}{J} - \left(\frac{h}{J}\right)_c\right]^{-\nu} \quad . \tag{5.27}$$

The recursion relation equations (5.23,24) are linearized when $h/J \sim (h/J)_c$,

$$\left(\frac{h}{J}\right)' - \left(\frac{h}{J}\right)_c = \lambda\left[\left(\frac{h}{J}\right) - \left(\frac{h}{J}\right)_c\right] \tag{5.28}$$

and $1' = 1(h/J)' = [1(h/J)]/2$ so that $1/\nu = \ln\lambda/\ln 2$. We get $\nu = 1.47$, compared with the exact value $\nu = 1$. In the transformation the lattice length spacing a goes to a' with

$$\frac{a'}{a} = n_s = 2 \quad .$$

The energy spacing is also changed, and $J'/J = h'/h$ is given by n_s^{-z} if z is the dynamical exponent (for the Ising model in a transverse field z = 1 so that time plays the role of an extra dimension).

At the nontrivial fixed point we have

$$z = \ln(J/J')/\ln(n_s) = \ln(h/h')/\ln(n_s) \quad . \tag{5.29}$$

Using (5.23,24) taken at the fixed point, we get z = 0.55 (the exact value is 1). The time rescaling is not done properly in the renormalization transformation.

From scaling arguments satisfied within this method, the gap G behaves like $G \sim [h/J - (h/J)_c]^s$ with $s = \nu z$. A third exponent can be calculated; the magnetic exponent β is defined by

$$M^x = \left[\left(\frac{h}{J}\right)_c - \frac{h}{J}\right]^\beta \quad . \tag{5.30}$$

At the fixed point we have

$$M^x = M^x\, n_s^{d_x}$$

with

$$n_s^{d_x} = \xi^{-1} = \frac{\sqrt{2}\sqrt{1 + a^2}}{1 + a} \tag{5.31}$$

taken at $(h/J)_c$. From scaling arguments $\beta = \nu d_x$. For $n_s = 2$ we get $\beta = 0.4$, compared with the exact value $\beta = 0.125$.

Better results can be obtained by going to next order in perturbation [5.46]. We still keep the two lowest levels of the block, and the higher levels now play a role through the perturbation scheme. The main change from the previous scheme is that the new Hamiltonian H' contains a new piece of the form

$$K \sum_i S_i^x S_{i+1}^z S_{i+2}^x \quad .$$

There are now four recursion relations for the couplings constants h, J, K and for the constant term E,

$$h' = \frac{E_1 - E_0}{2} + \frac{J^2}{4(1 + a^2)} (1 - a^2)^2 \left(\frac{1}{2E_0} + \frac{1}{2E_1} - \frac{1}{E_0}\right) \quad ,$$

$$J' = \frac{(1 + a)^2 J}{2(1 + a^2)} - 2K \frac{1 - a^2}{2(1 + a^2)} \quad ,$$

$$K' = -\frac{J^2}{8} \frac{(1 - a^2)^2}{(1 + a^2)} \left(\frac{1}{E_0} + \frac{1}{E_1}\right) \quad , \tag{5.32}$$

$$E' = 2E + \frac{E_1 + E_0}{2} + \frac{J^2}{8(1 + a^2)^2} \left[\frac{(1 - a)^4}{E_1 + E_0} + (1 - a^2)^2 \left(\frac{1}{E_1} + \frac{1}{E_0}\right)\right] \quad .$$

The analysis of these recursion relations leads to $(h/J)_c = 1.0985$, $\nu = 0.88$, and $z = 0.94$. The exponent β can be obtained by introducing an external field and looking at the recursion relations (new terms of the form $S_i^x S_{i+1}^z$ are also generated here in H') for the coupling constants. The value $\beta = 0.17$ is obtained. The method is thus greatly improved by going from first to second order in perturbation. The critical exponents are in much better agreement with the expected values; the dynamical exponent z has much improved and is close to the expected value $z = 1$. Calculations to higher order in perturbation are difficult because of the large number of new terms generated. This is shown here and by the case of the one-dimensional Hubbard model [5.84].

b) The Free Fermion Model in One Dimension

Let us consider the following free fermion model [5.37,84,88] on a lattice with the Hamiltonian

$$H = t \sum_{i=1}^{N-1} (c_i^+ c_{i+1} + c_{i+1}^+ c_i) + \sum_{i=1}^{N} (\mu\, C_i^+ C_i + E) \quad . \tag{5.33}$$

We shall first consider the half-filled band case $\mu = 0$. We divide the chain into blocks of n_s sites. The Hamiltonian is divided into intrablock and interblock parts.

The block Hamiltonian is solved exactly for general n_s. If we want to get H' with the same form as H with $\mu' = 0$, then we have to consider n_s odd blocks so that the lowest state is a doublet which is retained to form the new basis. When the method is applied to first order in perturbation the new Hamiltonian H' has the same form as H with new parameters t' and E',

$$H' = t' \sum_{j=1}^{N/n_s - 1} (c_j'^{+} c'_{j+1} + c'^{+}_{j+1} c'_j) + \sum_{j=1}^{N/n_s} E' \tag{5.34}$$

where t' and E' are functions of t and E. By solving (5.16) to first order in perturbation we get the recursion relations

$$t' = (-1)^l \frac{2}{(n_s + 1)} t \quad , \tag{5.35}$$

$$E' = n_s E + \left(\frac{1 - \cos \frac{\pi}{n_s+1} - \sin \frac{\pi}{n_s+1}}{1 - \cos \frac{\pi}{n_s+1}} \right) t \quad , \tag{5.36}$$

with $n_s = 2l + 1$.

Following the same analysis as in Sect.5.3.2a (5.29) the dynamical exponent z is given by

$$t'/t = n_s^{-z} = 2/(n_s + 1) \tag{5.37}$$

or

$$z = \ln[(n_s + 1)/2]/\ln(n_s) \quad .$$

For $n_s = 3$, $z = \ln 2/\ln 3 = 0.693$ and slowly tends toward the exact value $z = 1$ when n_s goes to infinity in a log fashion.

From the recursion relations (5.35,36) we find for the ground-state energy per site E_g for a very large system $N \to \infty$,

$$E_g = \left(\frac{1 - \cos \frac{\pi}{n_s+1} - \sin \frac{\pi}{n_s+1}}{1 - \cos \frac{\pi}{n_s+1}} \right) \left(\frac{n_s + 1}{n_s(n_s + 1) - 2} \right) t \quad . \tag{5.38}$$

For $n_s = 3$, $E_g/t = -2\sqrt{2}/5 = -0.56568$; and for $n_s = 5$, $E_g/t = -(3/14)(1 + \sqrt{3}) = -0.5854$, compared with the exact value $E_g/t = 2/\pi = -0.6366$.

If we suppose that the correlation function $\langle 0/c_i^+ c_{i+r}|0\rangle = g(r)$ behaves for large r as $\sim 1/r^\eta$ the exponent η can be calculated in the same way as the exponent β in Sect.5.3.2a. By rewriting the function g(r) in the new basis we get an equation

$$g(r) = Ag(r/n_s) \tag{5.39}$$

with $A = 2/(n_s + 1)$, and thus

$$\eta = \ln[(n_s + 1)/2]/\ln n_s = z \quad . \tag{5.40}$$

We also find evidence of oscillations in $g(r)$ as expected, and η goes to one when $n_s \to \infty$ as expected from the exact expression for $g(r) \sim 1/r \cos(\pi r/2)$.

The calculation can also be made to second order in perturbation. No new term appears in H', the recursion relation (5.35) is unchanged, and an extra term appears in (5.36) for E'. We did not calculate the recursion relation for general n_s but only for $n_s = 3$ and $n_s = 5$. For $n_s = 3$

$$E' = =E - \sqrt{2}t - \frac{5\sqrt{2}}{32}t \quad , \tag{5.41}$$

and for $n_s = 5$

$$E' = 5E - (1 + \sqrt{3})t - \{[12(1 + \sqrt{3})]^{-1} + (16\sqrt{3})^{-1} + 11/48\}t \quad . \tag{5.42}$$

From (5.41,42) together with (5.35) we get for the ground-state energy per site $E_g/t = -0.6541$ for $n_s = 3$ and $E_g/t = -0.6488$ for $n_s = 5$ which is very close (less than 3%) to the exact value.

We can also consider the general case with $\mu \neq 0$. There are three recursion relations. They carry an index i, integer between 1 and n_s, so that there are 3 n_s different recursion relations. Depending on the value of μ, the lowest levels retained are not the same and the recursion relations change. There are n_s unstable fixed points $(\mu/t)_i$. For each fixed point an exponent z and η can be defined. Two fixed points are different and correspond to the band edges where the "critical behavior" is different with $z = 2$ and $\nu = 1/2$ and where a gap opens linearly with μ. The recursion relations are now

$$\mu_i' = \mu_i - 2t_i \cos(i\pi/n_s + 1) \quad , \tag{5.43}$$

$$t_i' = [(-1)^{i+1}2/(n_s + 1)]\sin^2[i\pi/(n_s + 1)] \quad , \tag{5.44}$$

$$\begin{aligned} E_i' &= n_s E_i - \mu_i(i - 1) + 1 - \left(\frac{\cos(i - 1)x - \cos(ix)}{1 - \cos x}\right) \quad , \quad i < (n_s + 1)/2 \quad , \\ E_i' &= n_s E_i - \mu_i(i) + 1 - \left(\frac{\cos ix - \cos(i + 1)x}{1 - \cos x}\right) \quad , \quad i > (n_s + 1)/2 \quad . \end{aligned} \tag{5.45}$$

For n_s odd and $i = (n_s + 1)/2$ we recover the half-filled band behavior corresponding to the fixed point $(\mu/t)_i = 0$. The exponent

$$z_i = \ln\{(n_s + 1)/2 \sin^2[i\pi/(n_s + 1)]\}/\ln(n_s) \quad . \tag{5.46}$$

At the edge $z_1 = z_{n_s} \to 3$ when $n_s \to \infty$. We do not recover the correct result $z = 2$. This is due to the fact that by taking only the two lowest states even for very large blocks n_s we do not have a correct description of the density of states. More levels have to be taken into account. The ground-state energy can be calculated as a function of μ/t for all fixed point values,

$$(\mu/t)_i = 2\{\cos[i\pi/(n_s + 1)]\}/\{1 - n_s \sin^2[i\pi/(n_s + 1)]\} \quad . \tag{5.47}$$

This study of the free fermion model in one dimension is analogous to the study of the spin 1/2 XY model in one dimension [5.68]. This is due to the correspondence between the two models through the Jordan-Wigner transformation. There are however some differences for the correlation functions. The spin-spin correlation function $<S_i^x S_{i+r}^x>$ behaves like $1/r^{1/2}$ and $g(r)$ behaves like $1/r$.

5.3.3 Extensions of the Method

a) Extension to Large Blocks

The effect of increasing n_s has been shown on the free fermion model. A similary study has been made for the Ising model in a transverse field [5.45] for n_s up to $n_s = 7$. The accuracy increases slowly, and the dependence is often logarithmic. The calculations rapidly become very long, and for most problems it is difficult to go beyond $n_s = 7$ or 8. For instance, for the ITF the exponent β obtained from the renormalization of a mean block spin seems to vary like $\beta = \beta^* + A/\ln n_s$. The results converge slowly to the "exact" ($n_s \to \infty$) values (for instance, for $n_s = 7$ with the 1 dim ITF $z = 0.74$ and $\nu = 1.16$, compared with the exact values $z = \nu = 1$).

b) Extension by Increasing the Number n_L of Levels Retained

When n_L is large the new Hamiltonian H' does not belong to the same class as H so that the parameter space is much larger. For the 1d ITF, the block method has been applied first with $n_s = 2$ and $n_L = 4$, 6, 8 [5.65], and more recently with $n_s = 3$ and $n_L = 4$ leading to much better results ($\beta = 0.124$ and $z = 1.1$ as compared with the exact values $z = 1$ and $\beta = 0.125$). When several levels are retained, the choice is made not only from energy considerations but also from symmetry considerations (or matrix elements). In the study of the 1d Kondo lattice [5.82], the 8 levels retained are not always the lowest but have the proper symmetry and the strongest matrix elements.

c) Other Extensions

Variational schemes have been also introduced in the method [5.35], the block wave functions are parameterized, and the parameter is fixed so that the ground-state energy is the lowest. Such methods lead, however, to inconsistencies (some quantities come out better and others much worse). Decimation schemes [5.40] and self-dual methods [5.42,43,47] have also been introduced and applied to the 1d ITF. The self-dual method gives exactly the critical point position. An extension of the original self-dual method gives improved values for the critical exponents [5.58].

For the one-dimensional quantum Potts model, the expected first-order transition is missed because of the self-duality constraint [5.43].

5.4 Applications of the Block Method

The block renormalization group (BRG) method has been applied to spin, fermion, and coupled fermion spin systems and to the quantum version of statistical mechanics and field theory models. In this section we shall review these applications made mainly at T = 0.

5.4.1 Spin Systems

Only a finite number of quantum levels correspond to a spin operator. By retaining the right levels in the BRG method the new renormalized Hamiltonian H' can still be written in terms of spin operators.

a) The Spin 1/2 Ising Model in a Transverse Field (ITF)

This simple model was first studied as a test and also as the quantum version of the well-known classical Ising model. The method has been applied both in one dimension [5.35,40,45-48] and in higher dimensions [5.46,49-53], in the presence of an external imaginary field [5.54-56], in the presence of disorder [5.57,58], and with impurities [5.106]. All the extensions of the method have been tested: second-order perturbation [5.46,51], n_L large, n_S large, duality [5.47], and variational methods [5.35]. Frustration effects have been studied for the antiferromagnetic ITF model on a triangular lattice [5.50] where the ground state highly degenerate at zero field becomes nondegenerate when the transverse field exceeds a critical value.

b) The XY, Heisenberg Spin 1/2 Chain

The one-dimensional spin 1/2 XY model in a transverse Z field has been studied with the BRG method [5.68]. There is a critical field $H = H_c$; for $H > H_c$ a gap opens. This corresponds to merging from a band (full or empty) in a free fermion model. For $H < H_c$ several unstable fixed points are found which correspond to the oscillatory character of the spin-spin correlation function. In complete analogy with the free fermion model described in Sect.5.3.2b there are n_S such fixed points. The BRG method has been applied both with $n_L = 2$ and $n_L = 4$. The exponent η for the power law fall off of the correlation function is found close to the expected value $\eta = 1/2$.

The Heisenberg spin 1/2 chain was investigated very early with this method [5.75,89,90] as well as the dimerized Heisenberg model [5.73,74], the Heisenberg model with long-range interactions [5.76], the Ising-Heisenberg model [5.87], and disordered Heisenberg chains [5.77-79]. The essential singularity of the Ising-Heisenberg model close to the antiferromagnetic limit is not recovered.

c) The XY Model in a Z Field for d = 2, 3

The XY model in a Z field has been studied both in two and three dimensions [5.69] where the correspondence with free fermion fails. The following results have been found: no order in the ground state for d = 2; and order appears for d = 3. The correlation function exponent η has been estimated.

In d = 2 on a triangular lattice the antiferromagnetic XY model in a Z field h is "frustrated". This model was studied for the first time with the BRG method [5.70-72]. The critical field h_c is reduced because of the frustration effects. Such a model could be applied to two-dimensional adsorbed systems [5.71].

d) The Spin 1 XY Model with an Anisotropy Field for d = 1

Let us consider the system described by the Hamiltonian

$$H = J \sum_i (S_i^+ S_{i+1}^- + S_i^- S_{i+1}^+) + h \sum_i S_i^{z^2} \quad . \tag{5.48}$$

This model can be looked at as a higher spin extension of the spin 1/2 XY model in a Z field, but it behaves very differently because the spin is now an integer. This model can also be introduced as a truncated version of the quantum formalism in 1 + 1 dimensions of the d = 2 classical XY model. The BRG method has been applied to this Hamiltonian in the context of the Kondo necklace [5.81] with blocks of n_s = 2, 3 sites and by keeping the three lowest states n_L = 3 (singlet + doublet). A more recent study [5.107] has been done in the context of the truncated version of the Hamiltonian formulation of the d = 2 O(n) model (Sect.5.4.4a).

5.4.2 Fermion Systems

Fermions on a lattice can be studied with the BRG method. Two states are associated to a Fermion operator and the mapping operation can be done as shown in Sect.5.3.2b for d = 1 and for d > 1 [5.88]. Interacting fermions in d = 1 have been studied.

a) The d = 1 Hubbard Model [5.84,85]

Let us consider a system described by the lattice Hamiltonian

$$H = t \sum_{i\sigma} (c^+_{i,\sigma} c_{i+1,\sigma} + c^+_{i+1,\sigma} c_{i,\sigma}) + \sum_i U n_{i\uparrow} n_{i\downarrow} \tag{5.49}$$

Where $n_{i\sigma} = c^+_{i\sigma} c_{i\sigma}$, $c^+_{i\sigma}(c_{i\sigma})$ is the creation (destruction) operator for an electron of spin σ on site i, t is the transfer integral between adjacent sites, and U is the on-site Coulomb repulsion energy. This model has been introduced by HUBBARD [5.108] to treat correlations in narrow bands. Equation (5.49) corresponds to the half-filled band case with zero chemical potential. For a nonzero value of U a gap opens; there is a critical value $(U/t)_c = 0$ corresponding to a metal insulator transition. The gap opens like $(U/t)^{1/2} \exp(-A\, U/t)$ and thus remains small up to $(U/t) \sim 1$. This marginality is related to the nesting property of the one-dimensional model. It is interesting to know if this behavior can be obtained with the BRG method. This method has been applied mainly with $n_L = 4$ (two doublets) and for blocks of $n_s = 3$ sites, to first order in perturbation [5.84,85], and to second order [5.85]. The Hubbard Hamiltonian is diagonalized for a block of three sites; two doublets are retained (they are the lowest states only for small U/t) from symmetry considerations so that the new Hamiltonian H' has the same form as H. It can be found from the recursion relations for U, t and the constant term which is generated by two fixed points that $U/t = 0$ is marginally unstable and $U/t = \infty$ is stable, which correspond respectively to a metallic phase and an insulating phase. By integrating the recursion relations the ground-state energy is calculated. The gap is also obtained and is found to vary for $U/t \sim 0$ like $\exp(-7.4\, t^2/U^2)$, in slight disagreement with the exact solution. Going to large blocks (n_s large) does not improve this result. By extending the number of levels retained the method becomes numerically difficult to control. Instead the method has been extended to second order but still keeping two doublets. Going to second order in the intercell coupling consists of taking into account matrix elements which correspond to transitions from a state in this truncated basis to an intermediate state in which at least one of the cells is in a higher state and then back to a state in the truncated basis. Seven new interactions are generated: both nearest-neighbor and next-nearest-neighbor and three site interactions. The fixed point map remains unchanged. The fixed point $U/t = 0$ still corresponds to free fermions. The second-order result for the ground-state energy is in much better agreement with the exact theory than the first-order result. The second-order results for the gap are worse than the first-order one except for very small U/t where the gap behaves like $\Delta/t \sim (U/t)^{11}$ (the exponential behavior obtained in the first-order calculation is lost by going to second order).

b) Interacting Fermions in d = 1 [5.87]

Let us consider a system of interacting spinless fermions with Hamiltonian

$$H = \frac{t}{2} \sum_i (c_i^+ c_{i+1} + c_{i+1}^+ c_i) + G \sum_i n_i n_{i+1} - \mu \sum_i n_i \quad , \tag{5.50}$$

and let us take the simple electron-hole symmetric case $\mu = G$ with an average of 1/2 fermion per site (half-filled). This model presents a transition at $T = 0$ from a conducting phase when G/t is less than 1 to an insulating phase for $G/t \gg 1$. Analytical results have shown that the transition is of the "essential singularity" type with the gap opening like $\exp\{-A[(G/t) - 1]^{1/2}\}$. The BRG method has been applied by keeping $n_L = 2$ levels and for blocks of $n_s = 3, 5, 7$ sites. Only power law behavior is obtained for the gap, as for the isomorphic Ising-Heisenberg $S = 1/2$ model (Sect.5.4.1b).

c) One-Dimensional Model of f and d Electrons with Hybridization V and fd Interaction U_{fd}

The ground state and the gap properties of a two band model (f and d) with hybridization V and f-d interaction U_{f-d} are studied [5.115]. The model is represented by the Hamiltonian

$$H = E_d \sum_i d_i^+ d_i - t \sum_i (d_i^+ d_{i+1} + d_{i+1}^+ d_i)$$

$$+ E_f \sum_i f_i^+ f_i + V \sum_i (f_i^+ d_i + d_i^+ f_i) + U_{fd} \sum_i n_{d_i} n_{f_i} \tag{5.51}$$

where d_i^+ is the creation operator of d states with hopping matrix elements t, and f_i^+ the creation operator of a localized electron at site i. Only the half-filled band case $\langle n_f \rangle + \langle n_d \rangle = 1$ is studied. Blocks of $n_s = 3$ sites are used and $n_L = 4$ levels are retained (the two lowest states in the n_s particle subspace and the lowest states in the $n_s - 1$ and the $n_s + 1$ particle subspaces). When $U_{fd} = 0$ the exact hybridization two-band model is recovered, and for $U_{fd} \neq 0$ new results are obtained. The system is an insulator if half filled for $U_{fd} \to 0$, and the correlation length diverges when $U_{fd} \to 0$ like $\exp(t/U_{fd})$.

5.4.3 Spin-Fermion Systems: The Kondo Lattice in d = 1

The Kondo lattice model is an extension of the one-impurity Kondo model to a lattice of Kondo impurities well suited to represent concentrated rare-earth systems. The Kondo lattice Hamiltonian is written

$$H_{KL} = J \sum_i [2(c_{i\uparrow}^+ c_{i\downarrow} S_i^- + c_{i\downarrow}^+ c_{i\uparrow} S_i^+) + (c_{i\uparrow}^+ c_{i\uparrow} - c_{i\downarrow}^+ c_{i\downarrow}) S_i^z]$$

$$+ \frac{W}{2} \sum_i (c_{i\uparrow}^+ c_{i+1\uparrow} + c_{i\downarrow}^+ c_{i+1\downarrow} + \text{h.c}) - \varepsilon_\Gamma \sum_i (c_{i\uparrow}^+ c_{i\uparrow} + c_{i\downarrow}^+ c_{i\downarrow}) \tag{5.52}$$

where $c_{i\sigma}^+$ is the creation operator for a conduction electron (bandwidth W) of spin σ and where the f electrons are represented by 1/2 spins $\underline{S}_i(S_i^x S_i^y S_i^z)$. Another model, the Kondo necklace, was also introduced as a spin 1/2 analog of the d = 1 Kondo lattice model [5.64-65] and studied with the BRG. A BRG study of the Kondo lattice model has been made later [5.80-82] with blocks of $n_s = 3, 4$ sites and retaining $n_L = 8$ levels with the Kondo-like symmetry. For the half-filled case a gap opens for $J \neq 0$ like $\exp(-W/J)$ when J/W is small in analogy with the d = 1 Hubbard model [5.83].

5.4.4 Quantum Versions of Classical Statistical Mechanics in 1 + 1 Dimension

a) The O(n) Model

Following KOGUT [5.98] the d = 2 classical O(n) vector model can be transformed through the transfer matrix method to a quantum problem of coupled rotators on a chain. By truncating some of the high-energy levels, this Hamiltonian can be transformed into a system of coupled singlet and n degenerate multiplet; n = 1 corresponds to a singlet-singlet system (spin 1/2) and n = 2 to a singlet-doublet system (spin 1). The study of this truncated Hamiltonian has recently been done with the BRG method for blocks up to n_s = 9 sites and by retaining the two lowest states (singlet n-multiplet) [5.107]. The results have been obtained for all values of n including n = 0, n = -2, and n < -2. The value n = 0 is of interest because of the correspondence with polymerization (in the limit n → 0 the truncated Hamiltonian is in fact the true Hamiltonian representing the polymer statistics). The results are encouraging and, for instance, the exponent ν found for n = 0 is close to the expected value ν = 0.75 for polymers in d = 2.

b) The P(q) Potts Model

From the transfer matrix approach of the Potts model [5.109] the time-continuum quantum version of the q-states Potts model can be written as

$$H = \sum_{i=1}^{N} -\frac{\lambda}{q} \sum_{k=1}^{q-1} M_i^k M_{i+1}^{q-k} - h \sum_{i=1}^{N} R_i \tag{5.53}$$

where N is the number of sites of the d = 1 chain and with

$$R = \begin{bmatrix} +\,q-1 & & & 0 \\ & -1 & & \\ & & -1 & \\ 0 & & & \ddots \\ & & & & -1 \end{bmatrix} \quad \text{and} \quad M = \begin{bmatrix} 0 & 1 & 0 & \dots \\ 0 & 0 & 1 & \\ \vdots & & & \\ 1 & 0 & 0 & \dots \end{bmatrix} \quad . \tag{5.54}$$

The form of this Hamiltonian shows that in this representation the effect of the Potts coupling is to raise one spin and lower the neighboring spins. The Potts spins are however not real spins (raising the Potts "spin" when it is in its highest state $|q\rangle$ brings it to its lowest state $|1\rangle$). There are q states per site, a (q - 1) degenerate multiplet and a singlet (the Ising singlet-singlet system is recovered for q = 2 and the percolation problem corresponds to q → 1). The Hamiltonian (5.53) has been studied with the BRG method [5.42-44] with the blocks of n_s = 2 sites. New coupling terms are generated and a first-order transition is found when q exceeds a critical value q_c = 6.81 [5.43] (from analytical analysis a first-order transition is expected for $q > q_c = 4$).

c) Tricritical Point for Ising Systems in 1+1 Dimensions

The classical Blume-Emery-Griffiths (BEG) [5.109] model in two dimensions can be reduced to the study of the following quantum model in one dimension [5.60-63]:

$$H = h_1 \sum_i S_i^x + h_2 \sum_i S_i^y + h_3 \sum_i S_i^z$$

$$- K_1 \sum_i S_i^z S_{i+1}^z - K_2 \sum_i S_i^{z^2} S_{i+1}^{z^2} \qquad (5.55)$$

where S_i^x and S_i^z are the components of a spin 1 operator. This Hamiltonian has been studied using the BRG method for blocks of $n_s = 2, 3$ spins and retaining the three lowest levels at each stage. A tricritical point is found as expected and the agreement with results obtained by other methods seems encouraging [5.60].

5.4.5 Applications to Field Theory

a) The Thirring Model in One-Space and One-Time Dimension [5.37]

The lattice Thirring model on a lattice in 1 + 1 dimensions is very analogous to the Hubbard model [5.108]. The lattice Hamiltonian is

$$H = \sum_{j_1, j_2} i\delta'(j_1 - j_2)\left(b_{j_1}^+ b_{j_2} - d_{j_1} d_{j_2}^+\right) - g \sum_j (b_j^+ b_j d_j^+ d_j - 1)^2$$

with

$$\delta'(j) = \begin{cases} -\dfrac{(-1)^j}{j} & j \neq 0 \\ 0 & j = 0 \end{cases} \qquad (5.56)$$

and where $b_j^+ d_j^+ (b_j d_j)$ are fermion creation (destruction) operators.

As in Sect.5.4.2a the BRG method is applied with three sites blocks. A critical value $g = g_c = 1.1$ is found. For $g < g_c$ the Hamiltonian renormalizes to the free-field theory ($g = 0$), and for $g > g_c$ it renormalizes to the "Heisenberg antiferromagnetic" behavior. This result agrees with what is known about the continuum Thirring model [5.110].

b) The U(1) Goldstone Model in Two Dimensions [5.36]

The Hamiltonian formulation of the U(1) field theory on a lattice is represented by the following Hamiltonian,

$$H = -\sum_i \frac{\partial^2}{\partial \theta_i^2} - \lambda \sum_i \cos(\theta_i - \theta_{i+1}) \quad . \qquad (5.57)$$

This describes a set of rotors, and adjacent sites are coupled by a force proportional to the angle between the rotors. The model (5.57) is analyzed in the limit of large λ using the BRG method. For sufficiently large λ there is a gapless region so that there is a critical value λ_c at which the gap vanishes and remains zero for $\lambda > \lambda_c$. When the block renormalization group method is applied, some high oscillator states are frozen out. The new effective Hamiltonian is more complicated than (5.57) with four independent parameters.

c) Lattice Gauge Theories [5.39-41]

The BRG method has been applied to the Z(2) Gauge field theory in 2 + 1 [5.39,40] and in 3 + 1 dimensions [5.39]. The results are rather preliminary. The Z(2) Gauge theory is defined by the Hamiltonian H,

$$H = -\sum_{l} S^x(l) - \lambda \sum_{P} S^z(p^1)S^z(P^2)S^z(P^3)S^z(P^4) \quad , \qquad (5.58)$$

where $S^x(l)$,$S^z(l)$ are Pauli matrices (spin 1/2) corresponding to the link l of the lattice and where the second term is a sum over all plaquettes of the hypercubic lattice. In d = 2 a dual transformation can turn H into the ITF model. Thus for the Z(2) Gauge model in 2 + 1 dimensions the three-dimensional Ising critical exponents should be recovered. A method very similar to the BRG method is applied to a Hamiltonian equivalent to H (5.58) but which is only semilocally gauge invariant. The results for the critical exponents of the transition do not agree very well with the expected values for the dual ITF model. The exponent z which should be 1 is found to be equal to 1/2. The same calculations have been extended to d = 3 dimensions where a second-order transition is found in agreement with a conjecture of MIGDAL [5.111] but in contradiction with Monte Carlo calculations [5.112] which are in favor of a first-order transition.

5.5 Discussion

We shall discuss the BRG method in comparison with other real-space RG methods and suggest some extensions for the method and its applications. A comparison is also made with recently developed finite-size scaling methods.

5.5.1 When is the BRG More Suitable?

The BRG method based on Wilson's renormalization group ideas is *approximate* and *nonperturbative*. It is used to calculate *ground-state* and low-energy properties of a *quantum* system defined on a d-dimensional (finite or infinite) *lattice*. It allows handling large many-body systems for which exact solutions (analytical) have not yet been found, i.e., are too difficult. This method, as every Renormalization method, is well adapted to study the properties of a quantum transition by taking into account the quantum fluctuations. It helps to go beyond the mean-field-like methods (Hartree, Hartree-Fock). As shown in Sect.5.4, the BRG method can be applied to almost any quantum system defined with boson, fermion spin, rotator, matrice-like operators. This method is iterative so that computers can be easily used. The iterative mapping scheme $H \to H'$ adds some new physics by enlarging the Hamiltonian parameter space and embedding the original system in a wider class of quantum systems:

a) when the number n_L of levels retained is large the new Hamiltonian H' is defined as a function of new operators (high spin or $n_L \times n_L$ matrices) [5.64];

b) when n_L is small but the method is applied to second order in perturbation [5.46, 84]. In the q-state quantum Potts model [5.43] a new term is generated in H' even when the method is applied to first order and this term drives the first-order transition when $q > q_c$.

In classical systems static and dynamic properties are not treated in a consistent way because the dynamics sometimes has to be introduced in an artificial way (this is the case for the Ising model). In quantum systems the dynamics and statics are interrelated through quantum mechanics; close to a transition the critical dynamical properties depend on the dynamical exponent z (dispersion relation $\omega = k^z$) which can be obtained from the BRG method.

5.5.2 How to Control the Method?

In many-body quantum theory perturbative approaches have been used for long time. Such techniques are well controlled (use of diagrammatic rules); any order in perturbation can be obtained in principle, and limits come from complicated counting and integrals. Such methods are inadequate to predict catastrophic behavior (phase transition) and can only be used to give information on a transition which is supposed to exist (analysis of power law singularities from series expansions, Paddê approximents, etc.). Nonperturbative renormalization methods overcome this difficulty. These methods are, however, more artistic than technical in the sense that *the rules to operate* are not well defined and the method has to be combined with physical intuition and with other methods. There is some freedom in the way to apply the BRG method and we shall now give some advice gained by working with this method for several years.

a) The Division of the Lattice into Blocks

The choice of the blocks is such that the new lattice belong to the same class as the original lattice. There is always some freedom in that choice. We can play with the size n_s, the edge-size ratio, and the symmetry.

The Symmetry. This choice is rather important if we want the symmetry of the original Hamiltonian to be conserved in the scheme. For instance, as shown in Sect.5.3 for d = 1 fermion systems with zero chemical potential or for Heisenberg d = 1 chains in zero external field, *odd* blocks have to be chosen so that no chemical potential (external field) is generated.

The Surface-Volume Ratio. Edge effects have to be minimized in the method because of the approximations introduced by treating the interblock part in perturbation. In d = 1 the edge effects go down rapidly as n_s^{-1} when n_s increases. For higher dimension the surface is still very large for small blocks, and very large blocks should be used to minimize these surface effects. This constitutes limiting the method to high dimensions. However this surface effect goes down rapidly when the method is applied to higher order in perturbation.

The Size n_s. A comparison of the method with exact results (toy models) has shown that results become more accurate when n_s is larger. The convergence is however very slow (log dependence for Ising systems). The exact results are not fully recovered when $n_s \to \infty$: the ground-state energy is exactly recovered but not the correct singular behavior of the gap and of the correlation functions as illustrated for the d = 1 Hubbard model [5.84].

b) Which Levels to Retain for the Truncated Basis?

The correcteness of the ground state and low-excited states obtained depend largely on the choice of the levels retained for the block Hamiltonian: energy, degeneracy, nature (matrix elements).

The Number n_L of Levels. Usually n_L is kept small so that the new Hamiltonian H' is not too complicated and can be handled numerically. When n_L is small the other levels can be included indirectly in the scheme by going to higher order in perturbation.

Energy. A basic rule is to choose the lowest levels so that the true ground state can be constructed, but also a set of levels so that the new Hamiltonian H' is close to the original Hamiltonian H. As shown for the Hubbard model [5.84] in Sect.5.4.2a, the four levels retained (two doublets) for stability of the scheme (H' is still Hubbard-like) are the lowest for small U/t but not for large U/t (the same happens for the Kondo lattice [5.80-82]).

Nature of the States. The choice of the block states also depends on the fact that these states are well coupled through the interblock Hamiltonian. For instance, in the d = 1 Kondo lattice analysis [5.82] the eight states retained are not the lowest but are well coupled - which seems a better choice (this cannot be proved).

It is sometimes useful to transform the original Hamiltonian so that the method works better. This is the case of the quantum version of the O(2) field theory in 1 + 1 dimensions. As shown in Sect.5.4.4a, the truncated version of O(2) when analyzed with BRG does not show the expected fixed line which is recovered in Sect. 5.4.5b when the full rotator model is considered and a well-suited choice of retained levels is made.

5.5.3 What Has Been Done and What Are the Difficulties Encountered?

a) Quantum Properties at T = 0

The method has been applied mainly at T = 0 for spin and fermion systems in one dimension with some exploration of other quantum systems [gauge models, quantum Potts, quantum truncated O(n)] and to d > 1. Apart from testing well-known models, new interesting results have been obtained, as for the ITF in an imaginary field [5.54,55], disordered Heisenberg chains [5.77-79], Heisenberg model with long-range interactions [5.76], and quantum spin models on frustrated lattices [5.50,70]. In most cases the method is applied to first order in perturbation with n_s up to 9 and n_L up to 8. The method has been applied to disordered systems.

b) Quantum Properties at T ≠ 0

The BRG method has been applied at T ≠ 0 for low temperatures to the interesting physical situation of disordered Heisenberg chains [5.78] by taking into account high-energy levels in the block so that thermodynamic functions can be calculated in a very approximate way.

c) Difficulties

Some difficulties have been found when the method has been applied to a model which is already well known. Each time an essential singularity or a "fixed line" were expected the method could not give it [Hubbard model, O(2) truncated model, Heisenberg chain]. This comes from the fact that the correct choice of physically important operators was not made correctly.

Other difficulties come from the fact that the method is a real-space method and not adequate to continuously explore electronic band structures defined in k space. This is demonstrated in the study in Sect.5.3.2 of the free fermion model where only a few values (n_s) of the filling of the band can be explored with the RG scheme.

The numerical accuracy of the exponent values is not perfect and converges slowly with the size of the blocks. Some results are never right even for a very large block as illustrated for free fermions in Sect.5.3.2.

5.5.4 Comparison Between Different Methods

We shall now discuss other similar methods.

a) The Real-Space RG Methods for Classical Systems

This method as shown in Sect.5.2 can be applied to quantum systems but seems to be limited to spin 1/2 systems because of the transformation between new and old operators. This difficulty disappears in the BRG method by the freedom in the choice of the truncated basis and the introduction of new operators.

b) Finite-Size Scaling Methods [5.99,105]

Such methods well suited to study the critical behavior have some similarity through the scaling idea with the BRG method. The quantum properties of small finite systems obtained exactly or semiexactly are extrapolated toward the large system using scaling ideas already developed before the renormalization rush. If the transition is well known this method can give quite accurate quantitative results, but is dangerous in trying to understand a very new problem. This method is complementary to the BRG method and can be used in parallel with it. Contrary to the BRG there is much less freedom. This method applied to the truncated O(2) 1 + 1 model has been successful in recovering the expected fixed line [5.113]. It has been recently applied to a complicated fermion model (Anderson lattice) introduced to describe concentrated rare-earth compounds [5.114].

5.6 What to Do Next?

5.6.1 Improvement of the Method

a) New variational tricks can be found so that the freedom of the method is finally restricted by minimizing the ground-state energy (such direction is actually pursued at SLAC).

b) The method has not been sufficiently explored at finite temperature.

c) Monte Carlo methods could be used for quantum systems if combined with renormalization group.

5.6.2 Applications

We choose three examples of future applications:

a) High dimension calculations ($d = 2, 3$), for instance, the Hubbard model in two dimensions.

b) High-spin systems in one and higher dimensions.

c) Coupled spin chains.

The block renormalization group method (BRG) has been shown to be a very useful tool for looking at the low-temperature properties of quantum lattice systems in a nonperturbative way with applications in solid-state theory, field theory, and statistical mechanics.

References

5.1 M.E. Fisher: Rep. Prog. Phys. *37*, 617 (1967)
5.2 L. Kadanoff, W. Götze, D. Hambler, R. Hecht, E.A.S. Lewis, V.V. Polcianskas, M. Rayl, J. Swift, D.J. Kane: Rev. Mod. Phys. *39*, 395 (1967)
5.3 K. Wilson, J. Kogut: Phys. Rep. *12*C, 75 (1974)
5.4 L. Onsager: Phys. Rev. *65*, 117 (1944)
5.5 H. Stanley: Phys. Rev. *179*, 570 (1969)
5.6 P. Pfeuty: Thesis, Paris (1970)
5.7 R.B. Stinchombe: J. Phys. C*6*, 2459 (1973)
5.8 V.G. Vaks, A.I. Larkin: JETP Sov. Phys. *22*, 678 (1967)
5.9 R.J. Elliott, C. Wood: J. Phys. C*4*, 2359 (1971)
5.10 P. Pfeuty, R.J. Elliott: J. Phys. C*4*, 2370 (1971)
5.11 P. Pfeuty: Ann. Phys. N.Y. *57*, 79 (1970)
5.12 E. Riedel, F. Wegner: Z. Phys. *225*, 195 (1969)
5.13 P. Pfeuty: J. Phys. C*9*, 3993 (1976)
5.14 J.A. Hertz: Phys. Rev. B*14*, 1165 (1976)
5.15 A.P. Young: J. Phys. C*8*, L309 (1975)
5.16 I.D. Lawrie: J. Phys. C*11*, 1123 (1978)
5.17 I.D. Lawrie, M.E. Fisher: Proc. 23rd Conference on Magnetism and Magnetic Materials, Minneapolis, 1977
5.18 Th. Niemeijer, J.M.J. Van Leeuwen: Phys. Rev. Lett. *31*, 1411 (1973); Physica *16*, 425 (1974)
5.19 J. Rogiers, R. Dekeyser: Phys. Rev. B*13*, 4886 (1976)
5.20 D.D. Betts, M. Plischke: Can. J. Phys. *54*, 1553 (1976)
5.21 J. Rogiers, D.D. Betts: Physica *85*A, 553 (1976)
5.22 R. Dekeyser, M. Reynaert, M.H. Lee: Physica *86–88*B, 627 (1978)
5.23 A.L. Stella, F. Toigo: Phys. Rev. B*17*, 2343 (1978)
5.24 R. Dekeyser, M. Reynaert, A.L. Stella, F. Toigo: Phys. Rev. B*18*, 3486 (1978)
5.25 R.C. Brower, F. Kuttner, M. Naunberg, K. Subbarao: Phys. Rev. Lett. *38*, 1231 (1977)
5.26 Z. Friedman: Phys. Rev. Lett. *36*, 1326 (1976)
5.27 K. Subbarao: Phys. Rev. Lett. *37*, 1712 (1976)
5.28 G. Um: Phys. Rev. B*15*, 2736 (1977); B*17*, 3670 (1978)
5.29 A.L. Stella, F. Toigo: Physica *89*A, 175 (1977)
5.30 Z. Friedman: Phys. Rev. B*17*, 1429 (1978)
5.31 M. Suzuki, H. Takano: Phys. Lett. *69*A, 426 (1979)
5.32 K. Wilson: Rev. Mod. Phys. *47*, 773 (1975)
5.33 H.R. Krishna-Murthy, K.G. Wilson, J.W. Wilkins: Phys. Rev. Lett. *35*, 1101 (1975)
5.34 S.D. Drell, M. Weinstein, S. Yankielowicz: Phys. Rev. D*14*, 487 (1976)
5.35 S.D. Drell, M. Weinstein, S. Yankielowicz: Phys. Rev. D*16*, 1769 (1977)
5.36 S.D. Drell, M. Weinstein: Phy s. Rev. D*17*, 3203 (1978)
5.37 S.D. Drell, B. Svetitsky, M. Weinstein: Phys. Rev. D*17*, 523 (1978)
5.38 B. Svetitsky, S.D. Drell, H.R. Quinn, M. Weinstein: Phys. Rev. D*22*, 490 (1980)
5.39 D. Horn, S. Yankielowicz: Nucl. Phys. B*161*, 533 (1979)
5.40 E. Fradkin, S. Raby: Phys. Rev. D*20*, 2566 (1979)
5.41 H. Hamber: Phys. Rev. D*24*, 941 (1981)
5.42 D. Horn, M. Karliner, S. Yankielowicz: Nucl. Phys. B*170*, 467 (1980)
5.43 J. Solyom, P. Pfeuty: Phys. Rev. B*24*, 218 (1981)
5.44 B. Hu: Phys. Lett. *79*A, 6 (1980)
5.45 R. Jullien, P. Pfeuty, J.N. Fields, S. Doniach: Phys. Rev. B*18*, 3568 (1978)
5.46 J. Hirsch, G. Mazenko: Phys. Rev. B*19*, 2656 (1979)
5.47 A. Fernandez-Pacheco: Phys. Rev. D*19*, 3173 (1979)
5.48 R. Jullien, K.A. Penson, P. Pfeuty: J. Phys. Paris *12*, L237 (1979)
5.49 K. Penson, R. Jullien, P. Pfeuty: Phys. Rev. B*19*, 4653 (1979)
5.50 K. Penson, R. Jullien, P. Pfeuty: J. Phys. C*12*, 3667 (1979)
5.51 J. Hirsch: Phys. Rev. B*20*, 3907 (1979)
5.52 D.C. Mattis, J. Gallardo: J. Phys. C*13*, 2519 (1980)
5.53 B. Hu: Phys. Lett. *71*A, 83 (1979)
5.54 K. Uzelac, P. Pfeuty, R. Jullien: Phys. Rev. Lett. *43*, 805 (1979)

5.55 K. Uzelac, R. Jullien, P. Pfeuty: Phys. Rev. B*22*, 436 (1980)
5.56 R. Jullien, K. Uzelac, P. Pfeuty, P. Moussa: J. Phys. Paris *42*, 1075 (1981)
5.57 K. Uzelac, K.A. Penson, R. Jullien, P. Pfeuty: J. Phys. A*12*, L295 (1979)
5.58 K. Uzelac, R. Jullien, P. Pfeuty: J. Phys. A*13*, 3735 (1980)
5.59 K. Uzelac, R. Jullien, P. Pfeuty: J. Phys. A*14*, L17 (1981)
5.60 H. Hamber: Phys. Rev. B*21*, 3999 (1980)
5.61 B. Hu: Phys. Lett. *75*A, 372 (1980)
5.62 D. Boyanovsky, L. Masperi: Phys. Rev. D*21*, 1550 (1980)
5.63 Y. Gefen, Y. Imry, D. Mukamel: Unpublished
5.64 R. Jullien, J.N. Fields, S. Doniach: Phys. Rev. Lett. *38*, 1500 (1977)
5.65 R. Jullien, J.N. Fields, S. Doniach: Phys. Rev. B*16*, 4889 (1977)
5.66 R. Jullien, J.N. Fields, S. Doniach: Bull. Am. Phys. Soc. *22*, 291 (1977)
5.67 R. Jullien, P. Pfeuty, J.N. Fields, S. Doniach: J. Phys. Paris C5 *40*, 293 (1979)
5.68 R. Jullien, P. Pfeuty: Phys. Rev. B*19*, 4646 (1979)
5.69 K.A. Penson, R. Jullien, P. Pfeuty: Phys. Rev. B*22*, 380 (1980)
5.70 R. Jullien, K. Penson, P. Pfeuty, K. Uzelac: Phys. Rev. Lett. *44*, 1551 (1980)
5.71 K.A. Penson, R. Jullien, S. Doniach , P. Pfeuty: In *Ordering in Two Dimensions*, ed. by S.K. Sinha (North Holland, Amsterdam 1980)
5.72 K.A. Penson, R. Jullien, P. Pfeuty: Phys. Rev. B (to be published)
5.73 J.N. Fields: Phys. Rev. B*19*, 2637 (1979)
5.74 J.N. Fields, J.C. Bonner, H.W. Blöte: J. Appl. Phys. *50*, 1807 (1979)
5.75 J.M. Rabin: Phys. Rev. B*21*, 2027 (1980)
5.76 J.M. Rabin: Phys. Rev. B*22*, 2420 (1980)
5.77 J. Hirsch, J. José: J. Phys. C*13*, L53 (1980)
5.78 J. Hirsch, J. José: Phys. Rev. B*22*, 5339 (1980)
5.79 J. Hirsch: Phys. Rev. B*22*, 5355 (1980)
5.80 R. Jullien, P. Pfeuty, A.K. Bhattacharjee, B. Coqblin: J. Phys. Paris C*5*, 331 (1980)
5.81 R. Jullien, P. Pfeuty, A.K. Bhattacharjee, B. Coqblin: J. Appl. Phys. *50*, 7555 (1980)
5.82 R. Jullien, P. Pfeuty, B. Coqblin: Conference on Valence Instabilities, Santa Barbara, January, 1981
5.83 R. Jullien, P. Pfeuty: J. Phys. F*11*, 353 (1981)
5.84 C. Dasgupta, P. Pfeuty: J. Phys. C*14*, 717 (1981)
5.85 J. Hirsch: Phys. Rev. B*22*, 5259 (1980)
5.86 J.G. Estêve, A.F. Pacheco, A. Sequi: Phys. Rev. D*23*, 1853 (1981)
5.87 G. Spronken: Thesis, Grenoble (1981)
G. Spronken, R. Jullien, M. Avignon: Phys. Rev. B*24*, 5356 (1981)
5.88 P. Pfeuty, R. Jullien, K.A. Penson: Unpublished
5.89 H.P. Van de Braak, W.J. Caspers, C. de Lange, M.W.M. Willanse: Physica *87*A, 354 (1977)
5.90 W.J. Caspers: Phys. Rep. *63*, 223 (1980)
5.91 R.P. Feynmann, A.R. Hibbs: *Quantum Mechanics and Path Integrals* (McGraw-Hill, New York 1965)
5.92 R.J. Elliott, P. Pfeuty, C. Wood: Phys. Rev. Lett. *25*, 443 (1970)
5.93 J. Lajzerowicz, P. Pfeuty: J. Phys. Paris C*5*a, 193 (1971)
5.94 M. Suzuki: Phys. Lett. *34*A, 338 (1971)
5.95 M. Suzuki: Prog. Theor. Phys. *46*, 1337 (1971)
5.96 M. Suzuki: Prog. Theor. Phys. *56*, 1454 (1976)
5.97 E. Fradkin, L. Susskind: Phys. Rev. D*17*, 2637 (1978)
5.98 J. Kogut: Rev. Mod. Phys. *51*, 659 (1979)
5.99 C.J. Hamer, M. Barber: J. Phys. A*13*, L169 (1980)
5.100 C.J. Hamer, M. Barber: J. Phys. A*14*, 241 (1981)
5.101 C.J. Hamer, M. Barber: J. Phys. A*14*, 259 (1981)
5.102 H.H. Roomany, H.W. Wyld: Phys. Rev. D*21*, 3341 (1980)
5.103 H.H. Roomany, H.W. Wyld, L.E. Holloway: Phys. Rev. D*21*, 1557 (1980)
5.104 H.H. Roomany, H.W. Wyld: Phys. Rev. B*23*, 1357 (1981)
5.105 R.R. Dos Santos, L. Sneddon: Phys. Rev. B*23*, 3541 (1981)
5.106 K. Uzelac, R. Jullien, P. Pfeuty: J. Phys. A*14*, L17 (1981)

5.107 M. Kolb, R. Jullien, P. Pfeuty: Unpublished
5.108 J. Hubbard: Proc. R. Soc. London A*276*, 238 (1963)
5.109 M. Blume, V.J. Emery, R.B. Griffiths: Phys. Rev. A*4*, 1071 (1971)
5.110 W. Thirring: Ann. Phys. N.Y. *3*, 91 (1958)
5.111 A.A. Migdal: Zh. Eksp. Teor. Fiz. *69*, 810, 1457 (1975)
5.112 M. Creutz, L. Jacobs, C. Rebbi: Phys. Rev. Lett. *42*, 1390 (1979)
5.113 R. Jullien, P. Pfeuty: J. Phys. A*14*, 3111 (1981)
5.114 R. Jullien, R. Martin: Solid State Commun. (in press)
5.115 J.E. Hirsch, W. Hanke: Phys. Rev. B (in press)

6. Application of the Real-Space Renormalization to Adsorbed Systems

M. Schick

With 18 Figures

The application of real-space renormalization-group methods to the description of order-disorder transitions that occur in adsorbed systems is described. Preservation of symmetries of the ground states by the renormalization-group transformation is stressed. Two methods that ensure this are presented: the sublattice method and the prefacing transformation. The concept of the vacancy, often associated with the use of the latter method, is introduced and the simplicity that it brings to the description of first-order transitions is emphasized. As an example, the description of the ferromagnetic transition of the two-dimensional q-state Potts model is reviewed. Numerous applications of these methods to problems in statistical physics are noted.

6.1 Introduction

Order-disorder transitions occur in a large number of adsorbed systems, are easily monitored by a wide variety of techniques, and provide a unique area for the testing of our current understanding of critical phenomena. It is for these reasons that they have become the focus of much attention in recent years. These transitions occur on surfaces in which the adsorbate is found preferentially at sites provided by a substrate. At low temperatures, a subset of these sites is occupied preferentially, forming a sublattice whose presence is easily detected by scattering experiments.[1] As the temperature is raised a transition occurs to a disordered phase in which all sites are occupied with equal probability. The transition is manifested in scattering experiments by the disappearance of the Bragg reflections arising from the superlattice and in thermodynamic experiments by singularities in such quantities as the specific heat or compressibility.[2] The number of adsorbate-substrate systems that exhibit commensurate ordered phases is indicated by a recent catalogue of such systems that lists some 444 references [6.5].

---

1 Several papers concerning neutron or electron scattering can be found in [6.1].

2 The classic work on specific heat of helium adsorbed on graphite is [6.2]; specific heat measurements of krypton on graphite are reported in [6.3]; the pioneering adsorption isotherm studies of graphite are those of [6.4].

There are excellent reasons for employing position-space renormalization-group (PSRG) methods to describe order-disorder transitions. First, the system can be modeled rather well by a lattice gas because the adsorbate atoms are located preferentially at adsorption sites. Thus the Hamiltonian of the system is invariant under finite, not-infinitesimal translations exactly as in a lattice gas. This makes the problem considerably more tractable than ones containing a continuous symmetry. Furthermore, the lattice gas is isomorphic to the Ising model [6.6], which has been the object of extensive applications of PSRG methods.

Second, the bulk of the information desired about order-disorder transitions in experimental systems pertains to the phase diagram. Traditionally, model phase diagrams have been compared with experiment to give insight into the nature of observed phases and to obtain values of interaction constants. Such nonuniversal information is readily obtainable from PSRG calculations. Further, the effect of changes in the interactions, reflecting perhaps changes in the experimental situation, is easily extracted. Last, the systems are two dimensional, so that the calculations are technically not very difficult.

The essential element of an order-disorder transition is the reduction of symmetry in passing from the disordered to the ordered phase. The symmetry aspects of the problem were first fully exploited by LANDAU [6.7], and they play a central role in the construction of the Landau free energy. These same considerations must remain at the heart of the renormalization-group formulation of the problem. That these symmetry considerations are maintained in the momentum-space formulations of the renormalization group is clear from the use of the Ginzburg-Landau-Wilson Hamiltonian, which is essentially a local Landau free energy functional. Their equal importance in position-space formulations was first noted by VAN LEEUWEN [6.8], who emphasized that the lowered symmetry of an ordered phase must be preserved under renormalization.

Two simple methods have been employed to ensure that this is accomplished. The first was introduced by VAN LEEUWEN in the paper noted above and is illustrated in Fig.6.1. The idea is to choose cells in such a way that an ordered ground state of site spins (here the simple antiferromagnetic structure on the square lattice) is mapped to an identical configuration of cell spins. Clearly the choice of cells shown in Fig.6.1 achieves this. The calculation suggested by VAN LEEUWEN was carried out by NIENHUIS and NAUENBERG [6.9]. They obtained distinct fixed points describing the antiferromagnetic, as well as the ferromagnetic, transitions as had been predicted [6.8]. They also obtained a phase diagram for this model in a five-dimensional parameter space. A similar technique was employed by SCHICK et al. [6.10] in the first application of PSRG methods to adsorbed systems, the description of the transition to the $\sqrt{3}\times\sqrt{3}$ phase which occurs in helium adsorbed on graphite [6.2]. The ordered ground state of this system is shown in Fig.6.2. A discussion of this calculation follows in Sect.6.2.

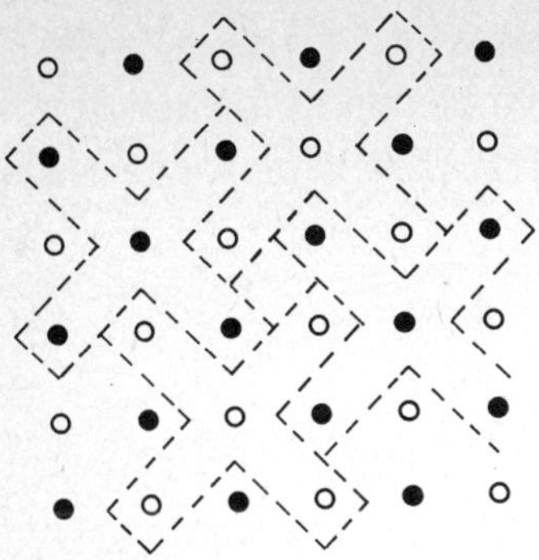

Fig.6.1. Decomposition of square antiferromagnet into two sublattices

Fig.6.2. Arrangement of atoms adsorbed on graphite in $\sqrt{3}\times\sqrt{3}$ structure. Shaded circles represent helium atoms

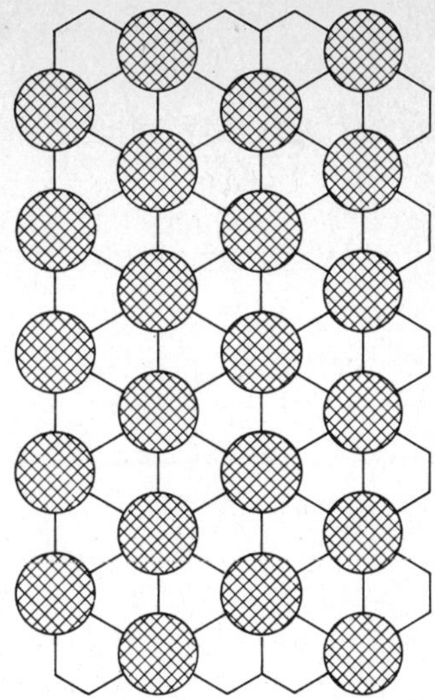

The second method employed to preserve the underlying symmetry of the transition is that introduced by BERKER et al. [6.11] and OSTLUND and BERKER [6.12] in papers describing the transition to the $\sqrt{3}\times\sqrt{3}$ phase exhibited by Kr on graphite [6.3,4,13]. The underlying symmetry of this order-disorder transition is identical to that of the ferromagnetic three-state Potts model [6.14,15], a fact first noted by ALEXANDER [6.16]. BERKER et al. argued therefore that the original system could be mapped to a ferromagnetic three-state Potts model. This model, being ferromagnetic, could then be dealt with more easily by standard methods. The properties of the original model then follow from the mapping. This method is discussed in Sect.6.3. As the calculations of Sects.6.2,3 apply to the same kind of transition, they afford a useful contrast of the two approaches.

A calculation of the phase diagram of a system undergoing a transition to a $\sqrt{3}\times\sqrt{3}$ phase using the method of BERKER et al. requires a reliable calculation of the phase diagram of the ferromagnetic three-state Potts model. A similar calculation of a system undergoing a 2×2 transition on a triangular lattice, a transition in the universality class of the four-state Potts model [6.17,18] requires a reliable calculation of the phase diagram of that model. Until recently, however, standard renormalization group calculations of Potts models were not considered reliable because they could not reproduce the result, known exactly, that the transition of the two-dimensional q-state Potts model is first order for q greater than four [6.19]. The reason for this failure and its rectification was shown by NIENHUIS et al. [6.20] to be again related to the necessity of preserving fundamental symmetries within the PSRG. For this reason the calculation is briefly reviewed in Sect.6.4. Further applications of the concepts used in Sects.6.3,4, including an application to adsorbed systems, is presented in Sect.6.5. We conclude with a brief summary.

6.2 The Sublattice Method

In 1971 BRETZ and DASH [6.21] observed a startling heat capacity signal arising from He^4 adsorbed on graphite at a density of 0.064 $Å^{-2}$. The data are shown in Fig.6.3. Because of the density of the adsorbate and the shape of the specific heat signal, they inferred that the helium had undergone a transition to the $\sqrt{3}\times\sqrt{3}$ structure shown in Fig.6.2. The interaction between two helium atoms adsorbed on nearest-neighbor sites provided by the graphite is repulsive due to the size of the helium atom (Lennard-Jones σ = 2.56 Å) and the distance between such sites (2.46 Å). Further-neighbor interactions are attractive. There are three equivalent ground states; this is made clear in Fig.6.4a, where the triangular lattice of adsorption sites is decomposed into three sublattices. In the ground states, one of the sublattices is occupied and the other two are empty. The transition is in the universality class of the three-state Potts model, which is also characterized by three equivalent ground states. Several groups have studied this transition by thermodynamic [6.2,22,23] and by scattering [6.24] methods, and much of the phase diagram has been mapped out. Thus there is a wealth of experimental data for comparison with a theoretical calculation.

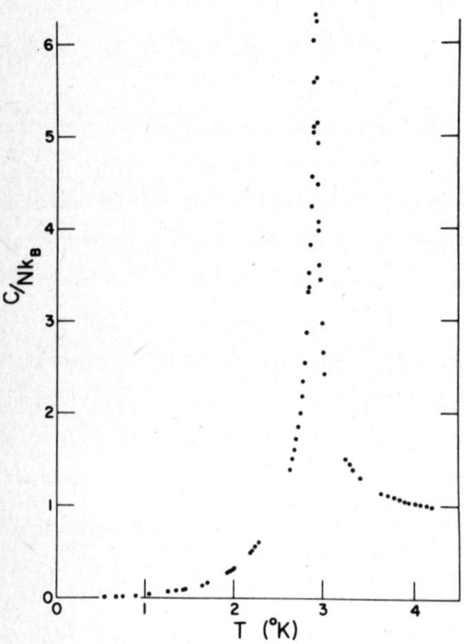

Fig.6.3. Specific heat of ^{4}He adsorbed on Grafoil at a density of 0.064 $Å^{-2}$

Most PSRG calculations begin with the assignment of cells, and that used by NIEMEIJER and VAN LEEUWEN [6.25] in their original treatment of the Ising ferromagnet and shown in Fig.6.4b will clearly not work here. As can be seen, the ground states, which are invariant only under a subset of lattice translations, are mapped in one step to a system invariant under any translation—i.e., to a fluid phase which

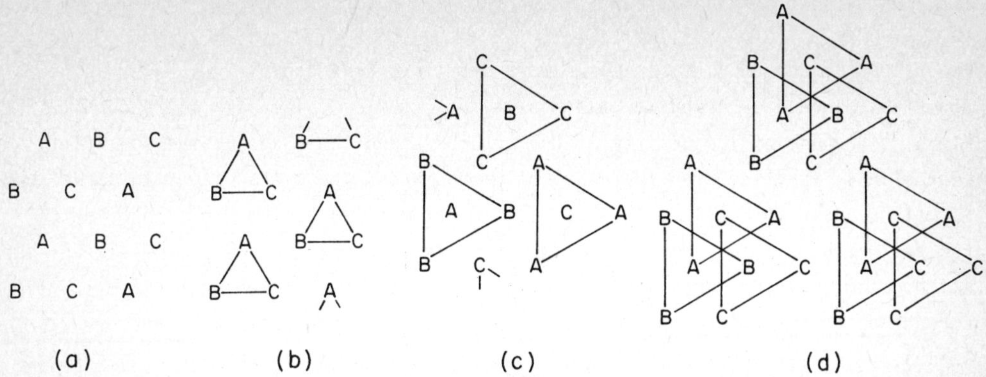

Fig.6.4. (a) Decomposition of triangular lattice into three sublattices; (b) Assignment of cells used in [6.11,12,25]; (c) An assignment of cells which preserves the $\sqrt{3}\times\sqrt{3}$ state used in [6.26,33]; (d) An assignment of cells which preserves the $\sqrt{3}\times\sqrt{3}$ state used in [6.10]

undergoes a liquid-gas transition. What is necessary is a cell assignment that preserves the sublattice symmetry. There are several possibilities, two of which are shown in Fig.6.4c,d. In the choice of Fig.6.4c, four lattice gas sites are mapped to one cell site. Of course there is the usual latitude in the choice of a projection operator, but the simplest is to ignore the central site and assign the cell occupation number according to whether a majority of the outer three sites are occupied or not. This scheme was applied by DEN NIJS et al. [6.26] to a different three sublattice problem, the triangular Ising model with ferromagnetic three-spin interactions solved by BAXTER and WU [6.27]. By employing eight cells and a cumulant expansion they obtained excellent results for the critical properties.

For the problem at hand, in which thermodynamic quantities are desired at all temperatures and not just in the critical region, a cumulant expansion is not appropriate; a finite lattice approach is indicated. Because each sublattice must be treated equally, a multiple of three cells is necessary. Further, the computational labor increases with the number of cells so that the choice of three cells is obvious. This has the consequence, however, that the only interactions that can be propagated by such a calculation are the nearest-neighbor interaction and a three-spin interaction. Thus the effect of attractive second - neighbor interactions is lost. This is certainly a limitation in many systems. However, in the helium system, there is a large portion of the phase diagram in which the repulsive nearest-neighbor interaction is dominant.

Much is known about the triangular lattice gas with nearest-neighbor repulsions. Such a model, with a chemical potential to control the density, is isomorphic to the triangular antiferromagnet in a magnetic field. The lattice gas is exactly soluble in two limits. First, at density one-half as the equivalent Ising model is in zero hence magnetic field and soluble [6.28]. The result is that the transition temperature is at zero degrees due to the lack of a well-defined ground state. The entropy per spin is finite, $S/N = 0.324$. That there is, in fact, a transition at zero temperature is indicated by the algebraic decay of correlations [6.29]. The universality class is that of the six-state clock model [6.18,30]. While of theoretical interest, these results are inapplicable to the helium system because at this high density the helium is not found in a commensurate ordered state in which the lattice gas model is appropriate. The model is also exactly soluble in the limit of infinite nearest-neighbor repulsion — the "hard-hexagon" model solved by BAXTER [6.31]. He finds a transition at the minimum ordering density $n_{min} = 0.276$. The critical exponent α is equal to 1/3, the same as that believed to characterize the three-state Potts model (Sect.6.4). From these results one can infer that for a system with finite nearest-neighbor repulsion, $T_c(n)$ will increase with n for $n > n_{min}$ and reach a maximum near $n \simeq 1/3$. It will then decrease to zero as the density approaches 1/2.

The simplest PSRG calculation that retains the sublattice symmetry involves three cells of three spins each as shown in Fig.6.4d. This calculation was carried out by SCHICK et al. [6.10], and the results are surprisingly good. In particular, they obtained $T_c(n = 1/2) = 0$, in agreement with the exact result, and an entropy per spin which differs by only 0.3% from the exact result. Further, a fixed point was found at $T = 0$ and thus a transition with power law decay of correlations. The minimum density of ordering was found to be 0.24, in reasonable agreement with the exact result of 0.276. One half of the phase diagram, which is symmetric about $n = 1/2$, is shown in Fig.6.5. It is compared there with a more recent phenomenological RG calculation [6.32] which I believe to be superior except near $n = 1/2$.

To compare this phase diagram with that of helium adsorbed on graphite one needs to fix a single parameter, the strength of the nearest-neighbor repulsion $4|J|$. This is done by fitting the maximum transition temperature, a single piece of data. Any thermodynamic function can now be calculated. The predicted variation of T_c with density is in satisfactory agreement with experiment, as shown in Fig.6.6. The range of densities over which the data exist, however, is insufficient to distinguish between the two phase diagrams shown in Fig.6.5. There are many specific heat data to compare with theory. Figures 6.7-10 show comparisons of the calculated specific heats at constant density with experiment for several densities. There are no adjustable parameters. The agreement for $T < T_c$ is extremely good, that for $T > T_c$ less so, presumably because the assumptions of the lattice gas are not as good; i.e., the atoms at high temperatures may find themselves at less favored intersitial sites or bridge sites not included in the lattice gas description. Further, the calculation yields a decrease of the amplitude of the singularity as the density changes from 1/3 exactly as observed in experiment. The specific heat in the critical region is poor, however, as the critical exponent obtained is $\alpha = -0.09$ instead of 1/3 expected theoretically or 0.36 measured experimentally by BRETZ [6.23]. With the exception of the critical region itself, this extremely simple calculation yields re-

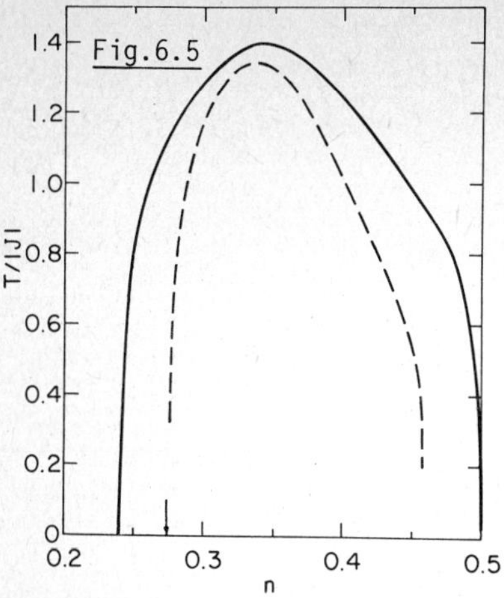

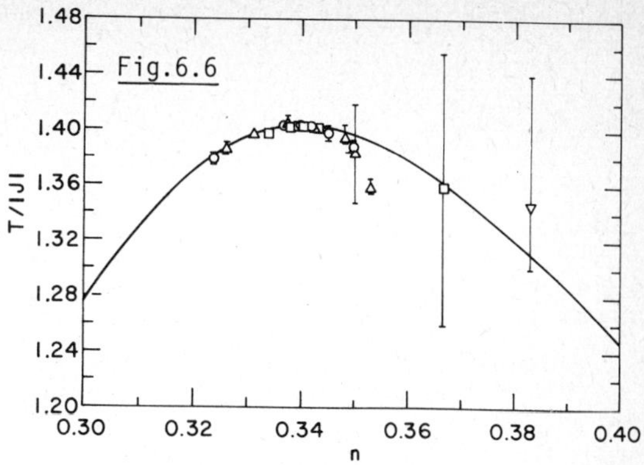

Fig.6.5. Phase diagrams of the triangular lattice gas with nearest-neighbor repulsion. Solid line, renormalization-group result of SCHICK et al. [6.10]; dashed line, phenomenological renormalization-group result [6.32]. The arrow is the exact result of BAXTER [6.31]

Fig.6.6. Phase diagram of SCHICK et al. [6.10] near the peak $n_{max} = 0.337$. Data points are for He^4 on Grafoil: circles, cell A [6.2]; triangles, cell B [6.2]; squares [6.23]; inverted triangles [6.22]

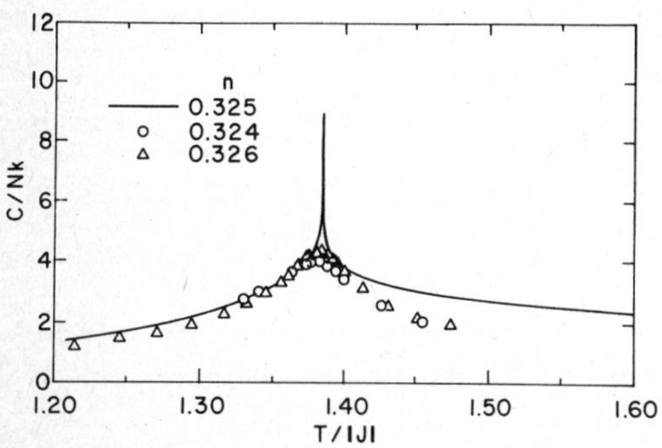

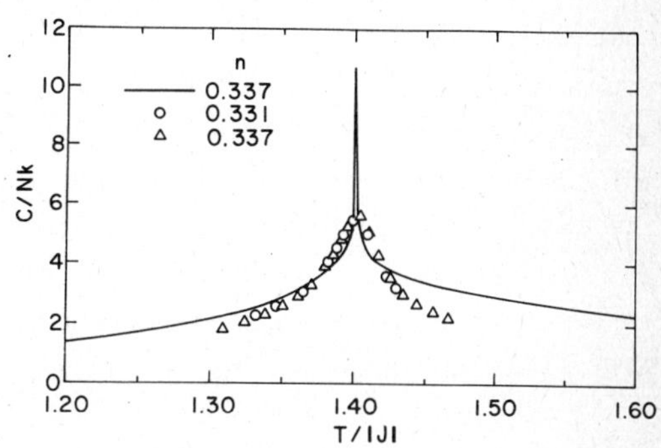

Fig.6.7. Specific heat at constant coverage. For densities somewhat below the maximum $n_{max} = 0.337$ [6.2]

Fig.6.8. Specific heat at constant coverage at n_{max}

sults in very good agreement with exact theoretical results and experimental observations.

A note of caution is appropriate here. It might be thought that a three-cell calculation employing the cells of Fig.6.4c would effect an improvement over the above as the center of these cells forms a uniformly spaced triangular lattice. Such a calculation, in fact, yields *no* transition whatsoever [6.33]. Why this particular assignment should fail so badly is unclear.

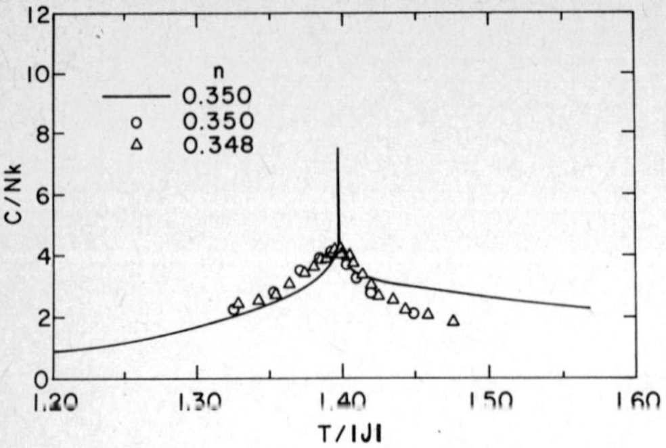

Fig.6.9. Specific heat at constant coverage. For densities somewhat above the maximum [6.2]

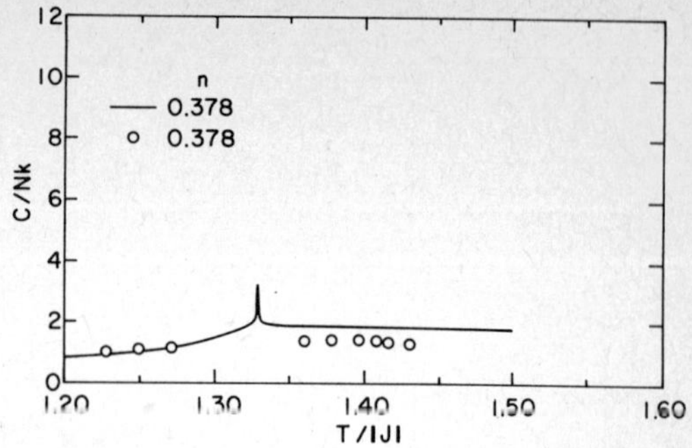

Fig.6.10. Specific heat at constant coverage. [6.4]

The sublattice method has been applied to several other problems of theoretical or experimental interest. These include transitions in various two- and three-dimensional Ising antiferromagnets [6.9,34,35], the antiferromagnetic Blume-Capel model [6.36], the antiferromagnetic three-state Potts model [6.37], two-dimensional bonded lattice fluids [6.38], lithium in $Li_xT_iS_2$ [6.39], and helium on krypton-plated graphite [6.40]. With the glaring exception noted above, there is little reason to doubt that these calculations provide rather good phase diagrams and poor critical exponents.

One advantage of this method is its simplicity. The recursion relations of the calculation described in detail above are easily programmed on a hand calculator. The few cells involved imply that only a few interaction constants are carried by the approximation. On the other hand, this is a weakness in that further-neighbor interactions cannot be handled without enormous increase in the size of the calculation. Further, if the system can undergo two different order-disorder transitons, the sublattice method can only describe it properly if the sublattice assignment is compatible with both symmetries, and this is usually not the case. Both of these problems are easily handled by the method discussed in the following section.

6.3 The Prefacing Method and Introduction of Vacancies

Let us consider again the triangular lattice gas with nearest-neighbor repulsive interactions. Other longer-range attractive interactions are not necessarily excluded. This model exhibits a transition to a $\sqrt{3}\times\sqrt{3}$ phase that by symmetry considerations, is in the universality class of the ferromagnetic three-state Potts model [6.16]. Therefore, the symmetry of the lattice gas transition can manifestly be preserved if a mapping from the lattice gas to a ferromagnetic three-state Potts model be affected. Such a mapping, called a prefacing transformation by BERKER et al. [6.11], who introduced the idea, can be written

$$\exp[H_{3p}(\{K_{3p}\}\{S_{3p}\})] = \sum_{\{n\}} P(\{S_{3p}\};\{n\}) \times \exp[H_{LG}(\{K_{LG}\};\{n\})] \quad , \tag{6.1}$$

where $\{K_{3p}\},\{S_{3p}\}$ are the set of couplings and statistical variables of the three-state Potts model and $\{K_{LG}\}$, n are the analogous set of the original lattice gas. The operator P is a suitably chosen projection operator. This transformation, carried out once, yields the three-state Potts couplings as a function of the original interaction

$$K_{3p} = K_{3p}(\{K_{LG}\}) \quad . \tag{6.2}$$

One now deals with a more tractable problem, a *ferromagnetic* three-state Potts model; thus the sublattice symmetry that made the original lattice-gas problem a difficult one is eliminated as the ground states of a ferromagnetic system are uniform. The price paid is only that the statistical variables are no longer those of an Ising model (n_i = 0, 1) but those of a three-state Potts model. This ferromagnetic model can now be treated by a variety of known methods. Once the critical surface of the three-state Potts problem is determined, that of the original system is obtained from the relations (6.2) of the prefacing transformation.

The prefacing transformation (6.1) differs from a renormalization-group transformation in that it maps a system with one kind of statistical variable onto another with a different kind of variable. It is employed only once. In practice, however, it is carried out in the same way as other PSRG transformations; e.g., cells and projection operators can be defined and the trace carried out on a system consisting of a finite number of cells. The cell choice of BERKER et al. is precisely that of Fig.6.4b, a choice which does not work if one maps the lattice gas to another lattice gas because, as noted earlier, it must yield a uniform ground state, which for a lattice gas is a fluid. It works here because one maps to a ferromagnetic Potts model that does indeed have uniform ground states.

Each cell of three sites has eight configurations that must be mapped to the Potts system. However, BERKER et al. simplified the calculation by assuming that the nearest-neighbor repulsion was infinite, an assumption that is reasonable for the larger gases like krypton (Lennard-Jones parameter σ = 3.61 Å), nitrogen (σ = 3.68 Å), and xenon (σ = 4.06 Å), all of which are observed to undergo a transition to the $\sqrt{3} \times \sqrt{3}$ phase [6.3,4,13,42,43]. Because of this infinite repulsion, the only possible cell configurations are those shown in Fig.6.11, and these are to be mapped to a three-state Potts system. The statistical variable of this system is a spin that can take three values S_i = A, B, C. BERKER et al. made the reasonable choice that the projection operator maps the cell configuration of Fig.6.11a completely to S_i = A, that of Fig.6.11b to S_i = B, and that of Fig.6.11c to S_i = C. Note that each of the three $\sqrt{3} \times \sqrt{3}$ ground states in the lattice gas are thereby mapped to the three ferromagnetic ground states of the Potts model.

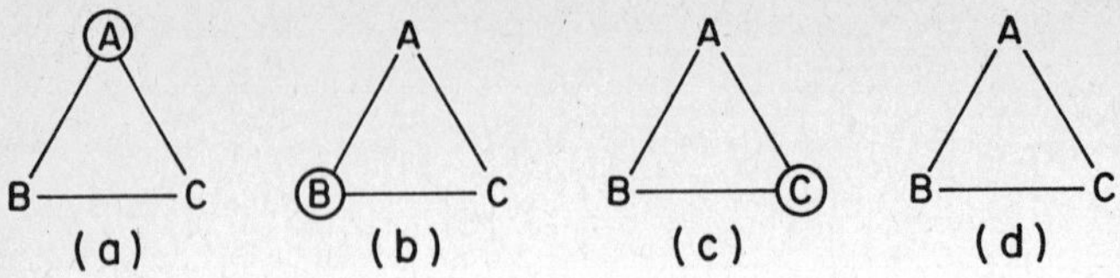

Fig.6.11. Allowed configurations of a cell with three sites when the nearest-neighbor repulsion is infinite

The question naturally arises as to the assignment of the configuration of Fig. 6.11d, in which the cell is empty. Such cell configurations will dominate when the original lattice gas is at low densities. It would be unwise to assign the cell with equal weight to the three possible states of the Potts spin, for in such an assignment the empty cell is viewed as encouraging the formation of the three ordered states equally when, in fact, it encourages none of them. BERKER et al. cleverly declared the site to be vacant: there would be no Potts spin on the site at all.

One of the consequence of introducing this possibility is that each site of the system to which the lattice gas is mapped must now be characterized by two variables: one, τ_i, which states whether the site is vacant ($\tau_i = 0$) or occupied by a Potts spin ($\tau_i = 1$); and, if the latter, the Potts spin itself, S_i. This generalized Potts model was denoted by BERKER et al. as a Potts lattice gas (PLG). For the case of nearest-neighbor interactions only, the Hamiltonian of this system is

$$H_{PLG} = \sum_{<ij>} \tau_i \tau_j (K + J\delta_{S_i S_j}) - \Delta \sum_i \tau_i \quad , \tag{6.3}$$

where Δ is a chemical potential governing the number of Potts spins, K is a lattice-gas interaction that if attractive, favors the occupation of neighboring sites by Potts spins, and J is the Potts interaction that favors the neighboring spins being in the same state.

This simple model has one feature that can be extraordinarily useful: within its parameter space there are first-order as well as continuous transitions. The continuous transitions are just those of the pure three-state Potts model, which remain continuous if the PLG is not too dilute. For large dilutions, however, (Δ large and positive), the transitions become first order just as in a liquid-gas transition not at its critical point. Thus the PLG is sufficiently complex to describe the non-universal features of the original adsorbed system, which also exhibits both first-order and continuous transitions.

Several remarks concerning the prefacing transformation are in order. First, as the mapping is now to a PLG, (6.1,2) are generalized to

$$\exp[H_{PLG}(\{K_{PLG}\};\{\tau, S_{3p}\})] = \sum_{\{n\}} P(\{\tau, S_{3p}\};\{n\}) \times \exp[H_{LG}(\{K_{LG}\};\{n\})]$$

$$K_{PLG} = K_{PLG}(\{K_{LG}\}) \quad . \tag{6.4}$$

Second, one can include interactions beyond the nearest neighbor in the original lattice gas. Of course this can be done in the first step of the sublattice procedure described in Sect.6.2. The crucial difference is that in the small subspace of the lattice-gas model used in that calculation, there are no first-order transitions. Therefore, attractive further-neighbor interactions cannot bring about a first-order transition in that calculation. The subspace of the PLG does contain such transitions, however. Attractive interactions in the original model tend to make Δ large and positive, the region of the PLG subspace where first-order transitions are encountered. Of course, such transitions in the PLG imply, via the inverse of (6.4), first-order transitions in the original lattice gas. The simplest prefacing transformation takes into account some or all of the second-, third-, and fourth-neighbor interactions.

Third, one need not confine oneself to a lattice-gas description of the adsorbed system. Instead, the cell can be thought of as consisting of the area surrounding three adsorption sites of the graphite, as shown in Fig.6.12. The Hamiltonian of the adsorbed system depends on the continuous coordinate of the atoms and contains two-body and one-body potentials. A projection operator maps the continuous system into the PLG. This is not new in the sense that any ordinary lattice-gas Hamiltonian describing an adsorbed system is obtainable in principle from a continuous Hamiltonian. The point here is that BERKER et al. did in fact begin with a continuous Hamiltonian containing experimentally determined interactions.

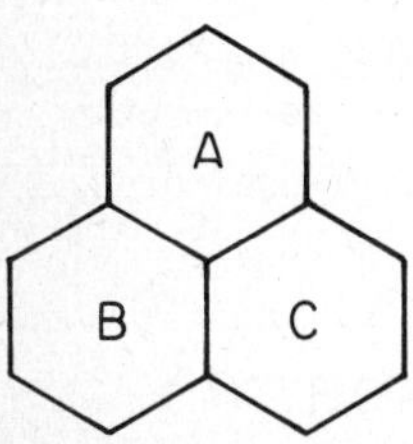

Fig.6.12. Portion of the graphite substrate that is mapped to three adsorption sites

Last, the prefacing transformation is carried out approximately, mapping a finite number of adsorption sites and adsorbed atoms onto a finite number of PLG variables. The number of interactions generated depends as usual on the number of independent cells. BERKER et al. used only two such cells in the prefacing transformation so that the Hamiltonian of the PLG contained only three couplings as in (6.3). Later, OSTLUND and BERKER [6.12] included three cells in the prefacing transformation, which enlarged the number of interactions in the PLG to twenty-two. In both calculations the resulting PLG was treated within a Migdal approximation scheme of their own devising.

The results of the latter calculation are shown in Fig.6.13 in solid lines. The ordered phase is to the left. There is a continuous transition to a fluid at high temperatures and a first-ordered transition at low temperatures. There is no critical point in the fluid phase and, therefore, no triple point. The calculation is

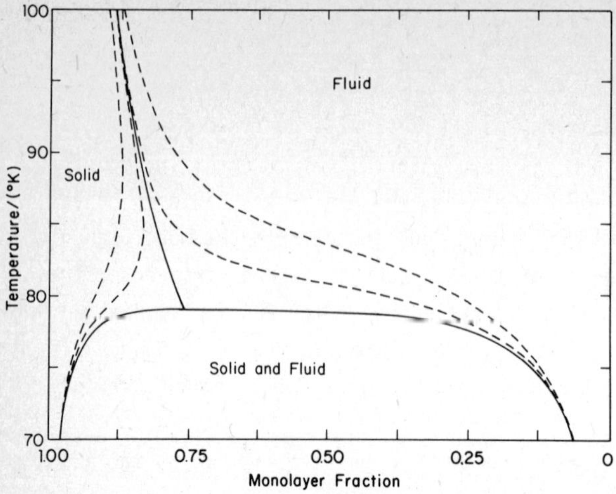

Fig.6.13. Calculated phase diagrams of krypton on graphite. Full curves correspond to an ideal thermodynamic system; dashed curves were calculated with temperature smearings of 0.1 and 0.8 K. A complete monolayer corresponds to the structure of Fig.6.2

quite capable of obtaining such a critical point and triple point, but none was obtained with the interactions used. As the existence of a critical point has been at issue [6.3,44], the calculation makes a significant contribution to the interpretation of experiment. The dotted lines show the smearing of phase boundaries brought about by model finite-size effects and inhomogeneities. Comparison with experimental data on Kr adsorbed on graphite [6.4,45,46] and N_2 adsorbed on graphite [6.42,44,47,48] are shown in Figs.6.14,15. The agreement on the first-order coexistence line, which is obtained with no adjustable parameters, is excellent.

As the above calculation abundantly demonstrates, the prefacing method can easily describe the occurrence of both first- and second-order transitions in adsorbed systems. A second advantage of the method is evident when a triangular system that can undergo transitions to either a $\sqrt{3}\times\sqrt{3}$ phase or a 2×2 phase is considered. Such a system has both first- and second-neighbor repulsive interactions. The transition to the 2×2 structure is in the universality class of the four-state Potts model [6.17,18] and has been studied by WALKER and SCHICK [6.49] employing the sublattice method. As noted in Sect.6.2, it is not possible in practice to consider two incompatible symmetries such as the above using this technique. However, it can be done using the prefacing transformation as shown by OSTLUND and BERKER [6.50]. A lattice-gas cell of twelve sites, compatible with both symmetries, is mapped to a three-state Potts variable if the occupation of the cell is like that of one of the $\sqrt{3}\times\sqrt{3}$ ground states, to a four-state Potts variable if the occupation is like a 2×2 ground state, and to a vacancy otherwise. The resulting Hamiltonian is treated within Migdal approximation. An extremely rich phase diagram results.

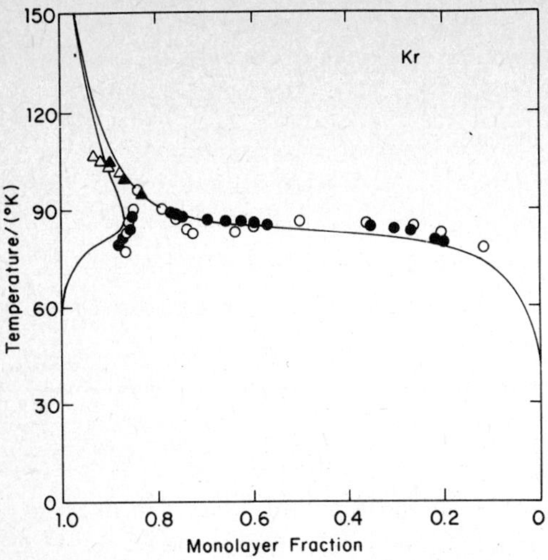

Fig.6.14. Predicted phase diagrams for krypton on graphite with temperature smearing of 0.8 K. Phase boundary points from experiment are as follows: first-order transitions from [6.4] (open circles) and [6.45] (closed circles); continuous transitions from [6.4] (open triangles) and [6.46] (closed triangles)

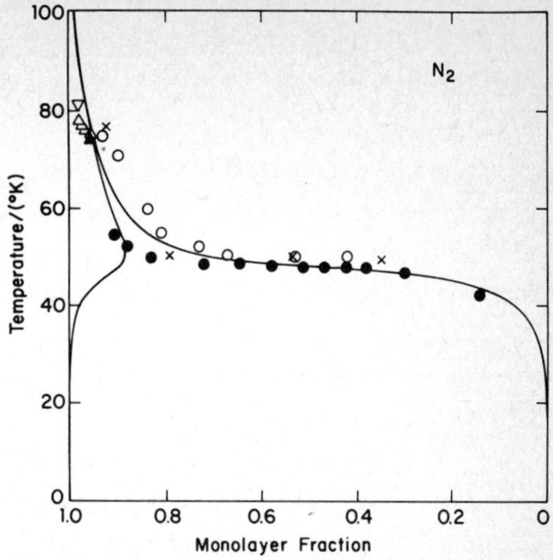

Fig.6.15. Predicted phase diagram for nitrogen on graphite with temperature smearing of 0.8 K. Phase boundary points from experiment are as follows: first-order transitions from [6.42] (crosses), [6.44] (closed circles), and [6.48] (open circles); continuous transitions from [6.47] (open triangles), [6.44] (closed triangle), and [6.48] (inverted triangle)

BERKER et al. [6.11] introduced two very useful concepts in their paper. The first, the prefacing transformation, has been seen to be of great utility and is certainly not restricted to problems in adsorption [6.51]. The second, the vacancy, is of even greater utility, a point which we shall demonstrate in Sect.6.4.

6.4 The Potts Model

The statistical variables of the Potts model are spins S_i, located at the ith site of a lattice, which can take on q values S_i = A, B, C, ..., q. The Hamiltonian of the model is invariant under the group of permutations of q objects. If there are nearest-neighbor interactions only, it takes the form

$$H = K \sum_{<ij>} \delta_{S_i,S_j} \quad . \tag{6.5}$$

The model has been of great theoretical interest because of the dependence of the nature of the transition to the ferromagnetic state upon the parameter q. Mean-field theory predicts the transition to be first order for all q greater than two due to the presence of a third-order invariant in the Landau free energy of the model. However, BAXTER [6.19] was able to prove that in two dimensions this prediction is only correct for q greater than four, at least for the nearest-neighbor model. For q less than or equal to four the transition is continuous.

This exact result led to many attempts to calculate the critical exponents of the two-dimensional Potts model. Because of the dimensionality of the problem, several calculations employed position-space methods and examined the transitions as a function of q. None of them was able to detect the change in the transition from continuous to first order [6.52]. Assuming no fundamental difficulties with the renormalization-group method itself, this failure must be attributable to the approximation methods employed. In retrospect the difficulty in describing the change in the nature of the transition can be formulated in terms of the difficulty of preserving symmetries.

As stressed earlier, a successful RG transformation must preserve the symmetries of the ordered states. In a first-order transition there are at least two ordered states which coexist, and the symmetries of both must be preserved. This is easy to do when dealing with the first-order transition in the ferromagnetic Ising model encountered when the magnetic field changes sign at low temperatures. The two coexisting ground states are fully ordered, one spin up, the other spin down. The symmetry is easily preserved and described by the renormalization group [6.53]. In the transition in the Potts model, however, only one of the coexisting phases is the fully ordered ferromagnetic phase. The other state is disordered, even at zero temperature, and has no broken symmetry. It remains invariant under the permutation of the q states.

A successful transformation must map every disordered state into a renormalized disordered state. The known approximation methods do not do this. Consider the cell method for example. If a disordered cell of Potts spins were to be mapped to a Potts cell spin in each of the q states with some probability, then the entire disordered configuration would map with some nonzero probability to each of the q ferromagnetic ground states, thus violating the symmetry.

NIENHUIS et al. [6.20] surmounted this difficulty by mapping such a disordered cell to a vacancy. This empty site in the renormalized system does not help neighboring Potts spins to align, just as on the original scale the disordered cell does not help its neighbor to align. Furthermore, a disordered configuration is mapped within a few steps to a lattice filled with vacancies. This representation of the disordered state is one whose symmetry is very easy to preserve. Therefore, a renormalization-group calculation employing the above idea easily preserves the symmetry of both phases which coexist at the first-order transition and thus is able to describe the change in the nature of the transition of the Potts model.

The first calculation of this kind [6.20] employed a two-cell approximation with three sites per cell on a triangular lattice. A generalized majority rule was employed; if all spins in a cell were in the same Potts state or two spins were in the same state and the third was vacant, then the cell was mapped to a cell spin in that Potts state. All other configurations were assigned to the vacancy. Thus, starting with a pure Potts model, vacancies were generated and the model became a Potts lattice gas as in Sect.6.3.

In the two-cell approximation only nearest-neighbor interactions are promulgated so that the Hamiltonian is

$$H_{PLG} = \sum_{<ij>} \tau_i\tau_j(K + J\delta_{S_iS_j}) - \Delta\sum\tau_i \quad . \tag{6.6}$$

As noted in Sect.6.3, the phase diagram of the model for q = 3 exhibits both continuous and first-order transitions. In terms of the renormalization group, there is, on the critical surface, a stable critical fixed point, an unstable tricritical fixed point, and a stable discontinuity fixed point. NIENHUIS et al. [6.20] found that flows from the pure Potts model subspace (Δ infinitely negative or K infinitely positive) are attracted by the critical fixed point for small q. Thus the transition is continuous. As q increases, the critical and tricritical fixed points approach one another and annihilate at a critical value q_c. Beyond q_c only the discontinuity fixed point remains to attract the flows, so that the transition becomes first order.

The topology of the critical surface is shown in Fig.6.16. Note that in the parameter space shown, the line EFG is a line of fixed points that is singly unstable (critical) from E to F and doubly unstable (tricritical) from F to G. The point F of the fixed line therefore plays a role analogous to the Kosterlitz-Thouless temperature on the fixed line in the Kosterlitz-Thouless description on the x-y model [6.54].

The numerical values of the critical value q_c and of the thermal exponents y(q) were not very accurate, as might be expected from such a calculation. For example, q_c was found to be 6.73, in contrast to the exact result of 4. However, the fact that tricritical and critical behaviors were part of the same fixed line indicated that thermal exponents for both kinds of transition would be given by the same function y(q). As DEN NIJS [6.55] had already conjectured that the critical exponent of the Potts model was related to that of the eight-vertex model y^{8V} by $(y - 3)(y^{8V} - 2) = 3$, with $y^{8V} = (2/\pi)\cos^{-1}(\sqrt{q}/2)$ taken along the branch $0 \leq y^{8V} \leq 1$. NIENHUIS et al. conjectured that the same relation must describe the tricritical exponents for $-1 \leq y^{8V} \leq 0$. These tricritical exponents can be measured in adsorbed systems. For example, the transition of krypton on graphite, which is in the universality class of the q = 3 Potts model, exhibits a tricritical point (Fig.6.14). From the above conjecture the thermal exponent y should cross over from its critical value of 6/5 to its tricritical value of 12/7. Similarly, the transition of helium on krypton-plated graphite, in the universality class of the Ising (q = 2 Potts) model [6.17], exhibits a tricritical point. The thermal exponent y should cross over from its critical value of 1 to the tricritical value 9/5.

Once it is seen that the Potts model can be imbedded in the larger subspace of the Potts lattice gas to describe its transition properly, any suitable approxi-

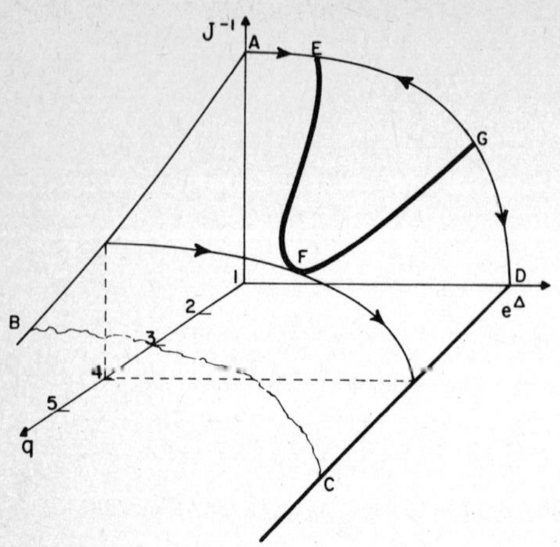

Fig.6.16. Phase diagram of the Potts lattice gas in the space of temperature, fugacity, and number of Potts states

mation can be applied to this more general model. NIENHUIS et al. applied the Kadanoff variational procedure, which is known to give excellent exponent values, and obtained both thermal [6.56] and magnetic [6.57] exponents in addition to a greatly improved value of $q_c = 4.08$. The thermal eigenvalues, obtained independently by BURKHARDT [6.58], were in excellent agreement with the DEN NIJS conjecture, as seen in Fig.6.17. This conjecture, at least for the critical exponents, has recently been derived by BLACK and EMERY [6.59].

The results of the calculation of the magnetic exponent, y_H, led NIENHUIS et al. [6.57] to conjecture that y_H is related to y^{8V} according to $(4y_H + y^{8V} - 6)(y^{8V} - 2) = -3$, where y^{8V} is given above. Again both critical and tricritical values of y_H are obtained according to the region of y^{8V} chosen. This conjecture, arrived at on different grounds by PEARSON [6.60], is strongly supported by a phenomenological renormalization-group calculation of BLÖTE et al. [6.61]. It reproduces all known exact results and predicts new ones.

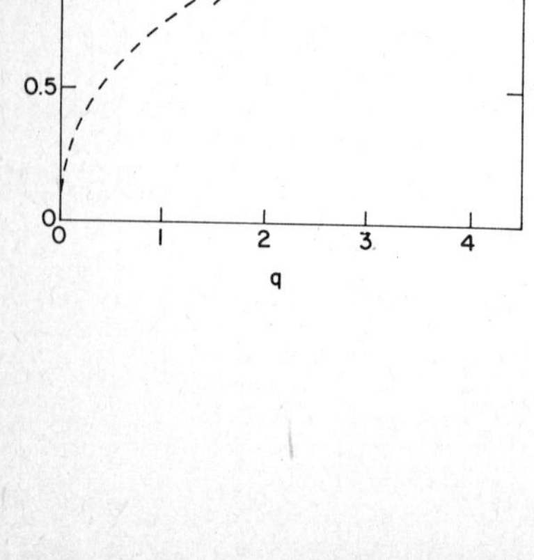

Fig.6.17. Comparison of the leading thermal exponent as calculated by NIENHUIS et al. [6.56] (solid line) with the DEN NIJS conjecture [6.55] (dashed line)

Of particular experimental interest is the value of β at the tricritical point of the system of krypton adsorbed on graphite. From the above two conjectures the value of $\beta = (2 - y_H)/y$ should, for q = 3, cross over from a critical value of 1/9 to a tricritical value of 1/18. Which of these two exponents should dominate near the tricritical point is determined by the cross-over exponent ϕ which is equal to the ratio of the second largest to the largest thermal exponent at the tricritical point. From the variational calculation [6.58], the second largest thermal exponent is 0.54. As the largest thermal exponent is conjectured to be 12/7, the value of ϕ is 0.31. This small value of ϕ indicates that a large experimental region in the vicinity of the tricritical point will be dominated by the tricritical behavior. Scattering experiments using synchrotron radiation as an intense source of X rays have been carried out on the krypton system near its tricritical point [6.62]. The measured value of β is 0.065 ± 0.015, in satisfactory agreement with the predicted $\beta = 1/18 = 0.056$. For the analogous case of a transition in the ferromagnetic Ising universality class, the exponent β is predicted to cross over from its critical value of 1/8 to its tricritical value of 1/24 with a cross-over exponent $\phi = 0.444$.

Additional information can be obtained from the topology of Fig.6.16 for the case of the four-state Potts model. According to this picture at the critical value q = 4, the stability of the fixed line changes, indicating a marginal operator. This idea can be incorporated into phenomenological scaling equations from which a rather complete picture of the singularity structure of the free energy at q = 4 can be extracted [6.63].

It seems clear that the thermal and magnetic exponents of the two-dimensional Potts model at both critical and tricritical points are now known exactly. A summary of these exponents is given in Table 6.1.

Table 6.1. Critical and tricritical exponents of the q-state Potts model. Other exponents can be obtained from scaling relations

Critical			Tricritical	
q	α	β	α	β
0	$-\infty$	1/6	1	0
1	-2/3	5/36	14/15	1/36
2	0	1/8	8/9	1/24
3	1/3	1/9	5/6	1/18
4	2/3	1/12	2/3	1/12

6.5 Further Applications of the Vacancy

NIENHUIS et al. [6.20] in their calculation broadened significantly the concept of the vacancy. They used it to stand for any region of local disorder or one without recognizable order of the kind associated with the underlying statistical variable.

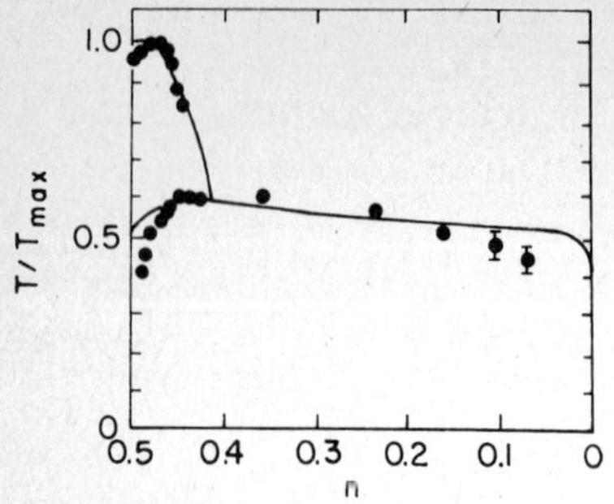

Fig.6.18. Comparison of the phase diagram of helium adsorbed on krypton-plated graphite as calculated by KINZEL et al. [6.40] and measured by TEJWANI et al. [6.65]

Whereas NIEMEIJER and VAN LEEUWEN assumed in their original work [6.25] that all regions of spins had to look like a spin itself, NIENHUIS et al. permitted a region to simply look disordered by assigning it to a vacancy. Seen in this light it is anticipated that the concept of the vacancy can be used with advantage in most PSRG calculations. Its introduction, of course, increases the space of Hamiltonians, thereby permitting more accurate results at the expense of increased labor. However, it decreases the labor of describing first-order transitions.

As an example of the introduction of the vacancy into an otherwise conventional PSRG calculation, consider again the sublattice method described in Sect.6.2. There the method was applied to an ordinary lattice gas with repulsive interactions, a model isomorphic to an antiferromagnetic Ising model in a magnetic field. If the sublattice cells are permitted to generate vacancies precisely as in the cell calculation of NIENHUIS et al. [6.20], the space of the antiferromagnetic Ising model in a field is enlarged to that of an antiferromagnetic Blume-Emery-Griffiths model [6.64] in a field. The vacancy can be thought of as equivalent to the $S_z = 0$ state of that $S = 1$ model. KINZEL et al. [6.40] employed this method to calculate the phase diagram of helium adsorbed on krypton-plated graphite. As there were two sublattices, two cells were chosen, one from each sublattice. Thus the Hamiltonian space included nearest-neighbor interactions only. It would be impossible to obtain first-order transitions in such a limited subspace of the original antiferromagnetic Ising model, but by introducing vacancies and passing to an antiferromagnetic Blume-Emery-Griffiths model, such transitions emerge naturally [6.36]. A comparison of the calculated phase diagram with experimental results [6.65] is shown in Fig.6.18.

Because of the ease of describing first-order transitions when vacancies are employed, they have been used in the calculation of the phase diagrams of several systems or models in which both first-order and continuous transitions are expected. Some examples are He^3-He^4 mixtures [6.66], the cubic model [6.67], the Potts model in arbitrary dimension [6.68,69], and the five-state clock model [6.70].

6.6 Summary

Position-space renormalization-group methods are ideally suited to the calculation of the phase diagrams of adsorbed systems. They are flexible and relatively simple. Two approximation methods have thus far been employed that preserve the symmetry of the ordered states, a sine qua non for a successful calculation. These are the sublattice method, with or without the introduction of vacancies, and the prefacing transformation. The results of these calculations have been extremely good and indicate that renormalization-group methods will become standard techniques in the calculations of surface physics.

Acknowledgments. Financial support from the National Science Foundation under grant DMR-7920785 is gratefully acknowledged. Alma Johnson prepared the typescript with her usual élan and efficiency.

References

6.1 J. Phys. (Paris) *38*, C-4 (1973)
6.2 M. Bretz, J.G. Dash, D.C. Hickernell, E.O. McLean, O.E. Vilches: Phys. Rev. A*8*, 1589 (1973); Erratum, Phys. Rev. A*9*, 2814 (1974)
6.3 D.M. Butler, J.A. Litzinger, G.A. Stewart, R.B. Griffiths: Phys. Rev. Lett. *42*, 1289 (1979);
D.M. Butler, J.A. Litzinger, G.A. Stewart: Phys. Rev. Lett. *44*, 466 (1980)
6.4 A. Thomy, X. Duval: J. Chem. Phys. *66*, 1966 (1969); *67*, 286 (1970); *67* , 1101 (1970)
6.5 G.A. Somorjai, M.A. Van Hove: *Adsorbed Monolayers on Solid Surfaces*, Structure and Bonding, Vol.38 (Springer, Berlin, Heidelberg, New York 1979)
6.6 M.E. Fisher: Rep. Prog. Phys. *30*, 615 (1967)
6.7 E.M. Lifshitz, L.P. Pitaevskii: *Statistical Physics*, 3rd ed. (Pergamon, Oxford 1980) Pt.1, Chap.14
6.8 J.M.J. van Leeuwen: Phys. Rev. Lett. *34*, 1056 (1975)
6.9 B. Nienhuis, M. Nauenberg: Phys. Rev. B*13*, 2021 (1976)
6.10 M. Schick, J.S. Walker, M. Wortis: Phys. Rev. B*16*, 2205 (1977)
6.11 A.N. Berker, S. Ostlund, F.A. Putnam: Phys. Rev. B*16*, 2205 (1977)
6.12 S. Ostlund, A.N. Berker: Phys. Rev. Lett. *42*, 843 (1979)
6.13 P.M. Horn, R.J. Birgeneau, P. Heiney, E.M. Hammonds: Phys. Rev. Lett. *41*, 961 (1978)
6.14 R.B. Potts: Proc. Cambridge Philos. Soc. *48*, 106 (1952)
6.15 F.Y. Wu: Rev. Mod. Phys. (to be published)
6.16 S. Alexander: Phys. Lett. A*54*, 353 (1975)
6.17 E. Domany, M. Schick, J.S. Walker: Phys. Rev. Lett. *38*, 1148 (1977)
6.18 E. Domany, M. Schick, J.S. Walker, R.B. Griffiths: Phys. Rev. B*18*, 2209 (1978)
6.19 R.J. Baxter: J. Phys. C*6*, L445 (1973)
6.20 B. Nienhuis, A.N. Berker, E.K. Riedel, M. Schick: Phys. Rev. Lett. *43*, 737 (1979)
6.21 M. Bretz, J.G. Dash: Phys. Rev. Lett. *27*, 647 (1971)
6.22 R.L. Elgin, D.L. Goodstein: Phys. Rev. A*9*, 2657 (1974);
S.V. Hering, S.W. Van Sciver, O.E. Vilches: J. Low Temp. Phys. *25*, 793 (1976)
6.23 M. Bretz: Phys. Rev. Lett. *38*, 501 (1977)
6.24 M. Nielson, J.P. McTague, W. Ellenson: J. Phys. (Paris) *38*, C4-10 (1977)
6.25 Th. Niemeijer, J.M.J. van Leeuwen: Phys. Rev. Lett. *31*, 1411 (1973); Physica (Utrecht) *71*, 17 (1974)
6.26 M.P.N. den Nijs, A.M.M. Pruisken, J.M.J. van Leeuwen: Physica (Utrecht) A*84*, 539 (1976)

6.27 R.J. Baxter, F.Y. Wu: Phys. Rev. Lett. *31*, 1924 (1973); Aust. J. Phys. *27*, 357 (1954)
6.28 R.M.F. Houtappel: Physica (Utrecht) *16*, 425 (1950);
G. Wannier: Phys. Rev. *79*, 357 (1950). Note that the numerical value of the entropy quoted by Wannier is incorrect. The correct value is given in our text
6.29 J. Stephenson: J. Math. Phys. *5*, 1009 (1964)
6.30 S. Alexander, P. Pincus: J. Phys. A*13*, 263 (1980)
6.31 R.J. Baxter: J. Phys. A*13*, L61 (1980)
6.32 W. Kinzel, M. Schick: Phys. Rev. B*23*, 3435 (1981)
6.33 K. Subbaswamy, G.D. Mahan: Phys. Rev. Lett. *37*, 642 (1976)
6.34 G.D. Mahan, F.H. Claro: Phys. Rev. B*16*, 1168 (1977)
6.35 B. Nienhuis, Aa.S. Sudbø, E.H. Hauge: Physica *92*A, 222 (1978)
6.36 G.D. Mahan, S.M. Girvin: Phys. Rev. B*17*, 4411 (1978)
6.37 M. Schick, R.B. Griffiths: J. Phys. A*10*, 2123 (1977)
6.38 A.P. Young, D.A. Lavis: J. Phys. A*12*, 229 (1979)
6.39 A.J. Berlinsky, W.G. Unruh, W.R. McKinnion, R.R. Haering: Solid State Commun. *31*, 135 (1979)
6.40 W. Kinzel, M. Schick, A.N. Berker: In *Ordering in Two Dimensions*, ed. by S. Sinha (Elsevier, Amsterdam 1980) p.381
6.41 M. Bretz: Ph. D. Thesis, Univ. of Washington (1971)
6.42 J.K. Kjems, L. Passell, H. Taub, J.G. Dash: Phys. Rev. Lett. *32*, 724 (1974);
J.K. Kjems, L. Passell, H. Taub, J.G. Dash, A.D. Novaco: Phys. Rev. B*13*, 1446 (1976)
6.43 J. Suzanne, J.P. Coulomb, M. Bienfait: Surf. Sci. *40*, 414 (1973)
6.44 T.T. Chung, J.G. Dash: Surf. Sci. *66*, 559 (1977)
6.45 Y. Larher: J. Chem. Soc. Faraday Trans. I*70*, 320 (1974)
6.46 F.A. Putnam, T. Fort, Jr.: J. Phys. Chem. *79*, 459 (1975), and *81*, 2164 (1977);
F.A. Putnam, T. Fort, Jr., R.B. Griffiths: J. Phys. Chem. *81*, 2171 (1977)
6.47 D.M. Butler, G.B. Huff, R.W. Togh, G.A. Stewart: Phys. Rev. Lett. *35*, 1718 (1975)
6.48 W.F. Brooks: Ph. D. Thesis, Stevens Institute of Technology (1977)
6.49 J.S. Walker, M. Schick: Phys. Rev. B*20*, 2088 (1979)
6.50 S. Ostlund, A.N. Berker: Phys. Rev. B*21*, 5410 (1980)
6.51 T.W. Burkhardt, W. Kinzel: Phys. Rev. B*20*, 4730 (1979)
6.52 C. Dasgupta: Phys. Rev. B*15*, 3460 (1977);
M.P.M. den Nijs: Physica (Utrecht) A*95*, 449 (1979), and references therein
6.53 B. Nienhuis, M. Nauenberg: Phys. Rev. Lett. *35*, 477 (1975)
6.54 J.M. Kosterlitz: J. Phys. C *7*, 1046 (1974)
6.55 M.P.M. den Nijs: J. Phys. A*12*, 1857 (1979)
6.56 B. Nienhuis, E.K. Riedel, M. Schick: J. Phys. A*13*, L31 (1980)
6.57 B. Nienhuis, E.K. Riedel, M. Schick: J. Phys. A*13*, L189 (1980)
6.58 T.W. Burkhardt: Z. Phys. B*39*, 159 (1980)
6.59 J.L. Black, V.J. Emery: Phys. Rev. B*23*, 429 (1981)
6.60 R.B. Pearson: Phys. Rev. B*22*, 2579 (1980)
6.61 H.W.J. Blöte, M.P. Nightingale, B. Derrida: J. Phys. A*14*, L45 (1981)
6.62 R.J. Birgeneau, G.S. Brown, P.M. Horn, D.E. Moncton, P.W. Stephens: J. Phys. C*14*, L49 (1981)
6.63 M. Nauenberg, D.J. Scalapino: Phys. Rev. Lett. *44*, 837 (1980);
J.L. Cardy, M. Nauenberg, D.J. Scalapino: Phys. Rev. B*22*, 2560 (1980);
L.P. Kadanoff: Phys. Rev. B*22*, 2560 (1980)
6.64 M. Blume, V.J. Emery, R.B. Griffiths: Phys. Rev. A*4*, 1071 (1971)
6.65 M.J. Tejwani, O. Ferreira, O.E. Vilches: Phys. Rev. Lett. *44*, 152 (1980)
6.66 A.N. Berker, D.R. Nelson: Phys. Rev. B*19*, 2488 (1979)
6.67 E.K. Riedel: Physica *106*A, 110 (1981);
B. Nienhuis, E.K. Riedel, M. Schick: To be published
6.68 B. Nienhuis, E.K. Riedel, M. Schick: Phys. Rev. B*23*, 6055 (1981)
6.69 A.N. Berker, D. Andleman, A. Aharony: J. Phys. A*13*, L413 (1980);
D. Andelman, A.N. Berker: J. Phys. A*14*, L91 (1981)
6.70 V. Mizrahi, E. Domany: Phys. Rev. B *24* , 4008 (1981)

7. Position-Space Renormalization Group for Models of Linear Polymers, Branched Polymers, and Gels

H. E. Stanley[1], P. J. Reynolds[2], S. Redner, and F. Family[3]

With 15 Figures

In this chapter we describe how position-space renormalization group methods can be used to treat models of linear polymers, branched polymers, and gels. Apart from their increasingly widespread use to describe phenomena occurring in polymeric systems, these models are of great interest in their own right, since they are purely geometric models embodying the full richness of ordinary thermal critical phenomena.

The organization of this chapter is as follows. Section 7.1 describes the three physical systems that are the principal focus of this review. In Sect.7.2, we describe three mathematical models that have proved useful in capturing the essential physics of the corresponding polymeric systems. Section 7.3 is devoted to a somewhat pedagogical description of how position-space renormalization group methods can be applied to each model. Finally, in Sects.7.4,5, we briefly describe additional approaches and draw some general conclusions.

Our aim is a self-contained exposition of some of the basic ideas rather than a comprehensive review of all work on this subject. Indeed, the application of modern methods of statistical mechanics to polymers is such a rapidly expanding field that it would be impossible to set higher goals for this chapter. For a recent authoritative monograph on polymer physics, see e.g., [7.1].

7.1 Three Physical Systems

7.1.1 Linear Polymers

In this section we introduce three basic classes of polymeric systems. Polymers are macromolecules which form by the chemical bonding of many monomeric constituents. A typical number for the "degree of polymerization" is about 10^4. On the other hand, the number of particles in a system (such as a fluid) exhibiting ther-

---

[1] John Simon Guggenheim Memorial Fellow, 1980- 1981.
[2] Also at MMRD, Lawrence Berkeley Lab, Univ. of California, Berkeley, CA 94720.
[3] Present address: Department of Physics, Center for Polymer Studies, Emory University, Atlanta, GA 30322. Supported in part by grants from ARO, ONR, and NSF.

mal critical phenomena is upwards of 10^{20}. Fortunately, it turns out that even for a polymer with "only" 10^4 monomers, statistical mechanics is a useful and highly appropriate tool.

Suppose, for simplicity, that all the monomers are identical, each having f functional groups which can react with one of the f groups of another monomer. The functionality is important in determining the topological structure of the resulting polymer. The simplest case, f = 0, produces no reactions at all; an example might be argon. The next simplest case, f = 1, results in dimers only. If f = 2, however, we are able to form unbranched linear polymers (Fig.7.1a).

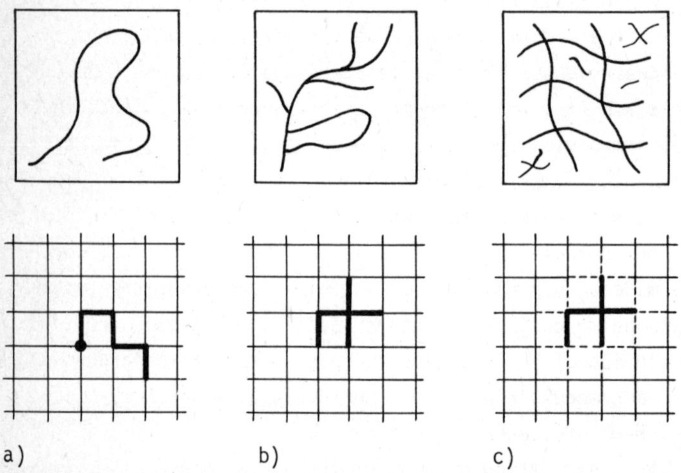

Fig.7.1a-c. Schematic illustration of the three polymeric systems and the lattice models used to describe their asymptotic or "critical" properties, (a) a linear polymer, modeled by an SAW, (b) a branched polymer, modeled by a random lattice animal, and (c) polyfunctional condensation (gelation), modeled by bond percolation. Specifically, (a) shows an L-step SAW whose global nonintersection constraint models the excluded volume effect for linear polymers in a good solvent. (b) shows an L-bond lattice animal, i.e., a connected L-bond cluster, which models a single branched polymer. This represents a generalization of an SAW in that branching can occur. Finally, (c) illustrates a single L-bond percolation cluster (actually many such clusters occur, cf. Fig.7.4). Such clusters are identical to lattice animals, but their relative probabilities of occurrence are different. All random animals with the same number of bonds are equally probable, while percolation clusters are weighted by an additional factor for each perimeter bond (----). In this figure, L is chosen to be 5 for the sake of simplicity only; in actual calculations, L might be as large as 10^2-10^6

Perhaps the most important physical feature of a polymer is the "excluded-volume" effect: the physical fact that two different monomers cannot occupy the same spatial position. For the case of a linear polymer this constraint means that the chain is non-Markovian in nature; hence it cannot be adequately described by a purely random-walk model.

One fundamental geometric property used to characterize a polymer chain in dilute solution is its mean end-to-end distance ξ. There has been considerable work in trying to understand the asymptotic dependence of ξ on the molecular weight M. This behavior is characterized by a critical exponent ν, defined by

$$\xi \sim M^{\nu} \quad [M \to \infty] \quad . \tag{7.1}$$

A simple approach for linear polymers, due originally to FLORY [7.2], has proved to be remarkably successful. This approach is based on minimizing the free energy of a polymer chain within a mean-field approximation. It leads to [7.3]

$$\nu_F = 3/(d + 2) \quad (d \leq 4) \quad , \tag{7.2}$$

where d is the spatial dimension of the system [7.1]. This simple formula is exact for d = 1 and d = 4, and appears to be a very accurate approximation for d = 2 and d = 3 as well. Above four dimensions, the Flory theory predicts that the chain is Gaussian, so that $\nu = 1/2$.

The intriguing simplicity and remarkable accuracy of (7.2) has stimulated considerable work on understanding the underlying basis of the Flory theory [7.1]. Although the physical arguments leading to (7.2) are mean field in character, the apparent reason for the high degree of accuracy is a delicate cancellation of errors in the energy and entropy terms in the Flory free energy [7.1,4]. However, an exact expansion for ν in the quantity (4-d) disagrees with (7.2) already at first order [7.5]. An expansion in (d-1), which may also be exact, also disagrees with (7.2) at first order, as we shall discuss in Sect.7.3.2.

Further interest in understanding linear polymers in dilute solution stems from the discovery of a correspondence between this system and the n-vector model of ferromagnetism [7.6] in the $n \to 0$ limit [7.5,7]. In this correspondence, it turns out that the variable M^{-1} in the polymer system corresponds to the variable $T - T_c$ in the magnetic system. Thus the limit of large molecular weight corresponds to a second-order phase transition. As a result, one expects quantities such as the polymer "diameter" to be characterized by universal behavior. In addition, the existence of a phase transition allows one to apply such modern techniques as the renormalization group to study the polymer problem.

7.1.2 Branched Polymers

The above remarks on linear polymers are readily generalized to more complicated structures. For example, if the constituent monomers have three or more functional groups ($f \geq 3$), we can form branched polymers as illustrated schematically in Fig. 7.1b.

Do we expect branched polymers to obey the same physical laws as linear polymers? Indeed the same "scaling law," (7.1), *is* obeyed for branched polymers, but the critical exponent ν is different.[1] The Flory theory for linear polymers can be gener-

Footnote 1 see next page

alized with the result [7.8-9]

$$\nu = 5/[2(d + 2)] \quad (d \le 8) \quad . \tag{7.3}$$

In contrast to the case of linear polymers, (7.3) gives the exact result for d = 3 [7.10] and d = 8 [7.11,12], and seems very accurate in d = 2. Moreover, while the magnetic analog of the linear polymer is the n → 0 limit of the n-vector model, the magnetic analog of the branched polymer is somewhat more complicated [7.11,13-15]. Table 7.1 compares the basic scaling properties of linear and branched polymers.

Table 7.1. Comparison of some of the "scaling" properties of (a) linear polymers, (b) branched polymers, and (c) gels. The first line gives d_c, the upper marginal dimension. The second line gives the critical exponent ν for $d \ge d_c$. The third and fourth lines give the Flory formulae for ν and their expansions in $\varepsilon \equiv d_c - d$ to lowest order. The fifth line gives the RG expansions, emphasizing the lack of agreement between ν_F and ν_{RG} in the immediate vicinity of d_c. Note that for the gel, ν is usually denoted ρ; it gives the dependence of the mean cluster diameter on the number of monomers in the cluster. Finally, we give the corresponding values of D^+, the "fractal" dimension ([7.16] and references therein). We note that D^+ is the same for branched polymers and gels above d_c, even though the numerical values of d_c differ

	(a) linear polymer	(b)branched polymer	(c) gel
d_c	4	8	6
$\nu(d \ge d_c)$	1/2	1/4	1/4
$\nu_F(d \le d_c)$	$3/(d+2)^a$	$5/[2(d+2)]^b$	$2/(d+2)^c$
$\nu_F(d \lesssim d_c)$	$\frac{1}{2}(1 + \frac{1}{6}\varepsilon)$	$\frac{1}{4}(1 + \frac{1}{10}\varepsilon)$	$\frac{1}{4}(1 + \frac{1}{8}\varepsilon)$
$\nu_{RG}(d \lesssim d_c)$	$\frac{1}{2}(1 + \frac{1}{8}\varepsilon)^d$	$\frac{1}{4}(1 + \frac{1}{9}\varepsilon)^e$	$\frac{1}{4}(1 + \frac{5}{42}\varepsilon)^f$
$D^+(d \ge d_c)$	2	4	4
$D^+(d \lesssim d_c)$	$2(1 - \frac{1}{8}\varepsilon)$	$4(1 - \frac{1}{9}\varepsilon)$	$4(1 - \frac{5}{42}\varepsilon)$

[a] [7.2], [b] [7.8,9], [c] [7.1], [d] [7.5], [e] [7.11], [f] [7.17-19]

7.1.3 Gels

Let us return to our simple example, the reaction of f-functional monomers (Fig. 7.1c). Figure 7.2 illustrates a specific case, trimethoyl benzene, with f = 3.

1 Specifically, ν is decreased. The slower increase in the linear diameter of the polymer when branching is allowed may be understood physically by realizing that monomers which form branches often do not contribute to an increase in the linear diamter ξ. The more branches that form, the more the number of additional branches that *can* form on the next step. This "cascade effect" causes a change in the *exponent* as opposed to the *amplitude* in (7.1).

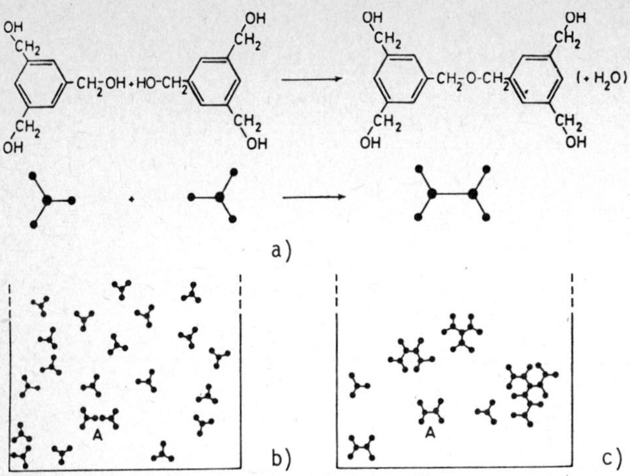

Fig.7.2a-c. Illustration of the simplest gelation phenomenon, polyfunctional condensation of f-functional monomers. The f-functional monomer shown in (a) is trimethoyl benzene; it has three "functional" groups that can react to form ether linkages. If f were two, then the most complex structures possible would be chains and rings; however, since f > 2 here, branched networks form. In (b) and (c) beakers are shown at successive stages of reaction. Adapted from [7.20]

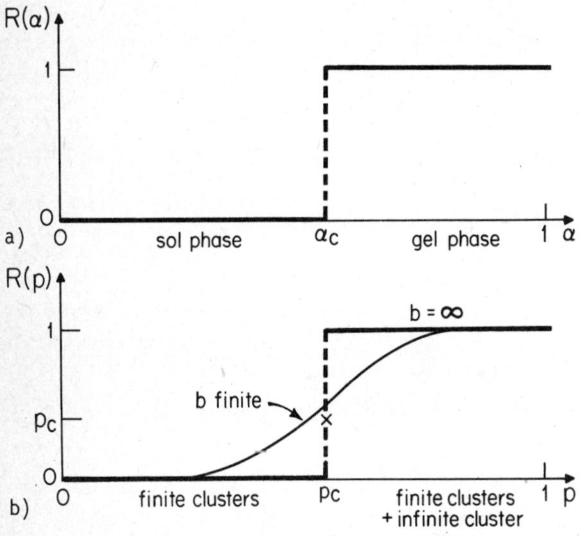

Fig.7.3a,b. Illustration of the analogy between (a) the gelation threshold and (b) the percolation threshold. In (a) the function R(α) denotes the probability that in an infinite system there is at least one molecule infinite in spatial extent. In (b), R(p) denotes the probability that there is at least one cluster infinite in spatial extent. The step function is for an *infinite* lattice, while the curve denotes the same probability function for a *finite* lattice of edge b (for which "infinite cluster" is interpreted to mean "spanning cluster"). The rounding apparent for the finite system becomes successively less pronounced as the system grows in size. In the limit of an infinite system, $R(p_c)=p_c$, so that p_c is a fixed point and the system is scale invariant. Thus in the renormalization group approach, percolation phenomena naturally divide into three regimes ($p<p_c$, $p=p_c$, and $p>p_c$)

Each benzene ring has three reactive methyl groups; methyl groups from two different monomers can react to form an ether linkage [7.20]. The reaction is characterized by a parameter α, termed the *conversion*, which is the fraction of reacted methyl groups. If $\alpha = 0$, only monomers are present. If $\alpha > 0$, finite polymers can exist in all sizes. However the probability of forming an infinite polymer molecule is zero for all α less than a critical value α_c. For $\alpha > \alpha_c$ the probability for an infinite molecule to occur jumps from zero to unity. Thus the *connectivity* of the system changes drastically at $\alpha = \alpha_c$, and this "phase transition" is termed gelation (Fig.7.3a).

7.2 Three Mathematical Models

In the study of polymer statistics, it is often useful to treat lattice models in which there is a 1:1 correspondence between monomers and sites of a lattice. The f reactive end groups may be thought of as directed along the f lattice bonds emanating from each lattice site. It is widely believed that such models display the same critical behavior as continuum systems ([7.21,22] and references therein).

In this section, we shall formally define the three lattice statistical models that are relevant to the three polymeric systems discussed above. For pedagogical clarity, we begin with a simple model used to describe gelation—percolation on a Cayley tree pseudolattice [7.23-29].

7.2.1 Percolation

The first successful model to capture the essential physics of the gelation threshold was proposed 40 years ago by FLORY, and developed in a series of classic papers by both FLORY [7.23] and STOCKMAYER [7.24]. The Flory-Stockmayer (FS) model not only predicts the existence of a gelation threshold α_c, but also provides the "critical exponents" characterizing the behavior of various quantities such as the gel fraction, degree of polymerization, and mean end-to-end distance ξ.

The FS model was formulated in a fashion that at first sight seems to be lattice independent. One merely requires that a given polymer be forbidden to loop back upon itself, i.e., intramolecular interactions are excluded. Today we recognize this assumption as fully equivalent (as far as critical behavior is concerned) to the statement that the polyfunctional monomers be required to occupy the sites of a Cayley tree: to each polymer configuration formed from bonding of f-functional monomers, there is a one-to-one correspondence with a configuration of bonds on a Cayley tree whose coordination number is equal to f [7.25-28].

In reality, loops do form—both by inter- and by intra-molecular bonding. What is the effect of allowing for the loops? Clearly the threshold will increase, since the extra bonds necessary to create the loops do not contribute to the for-

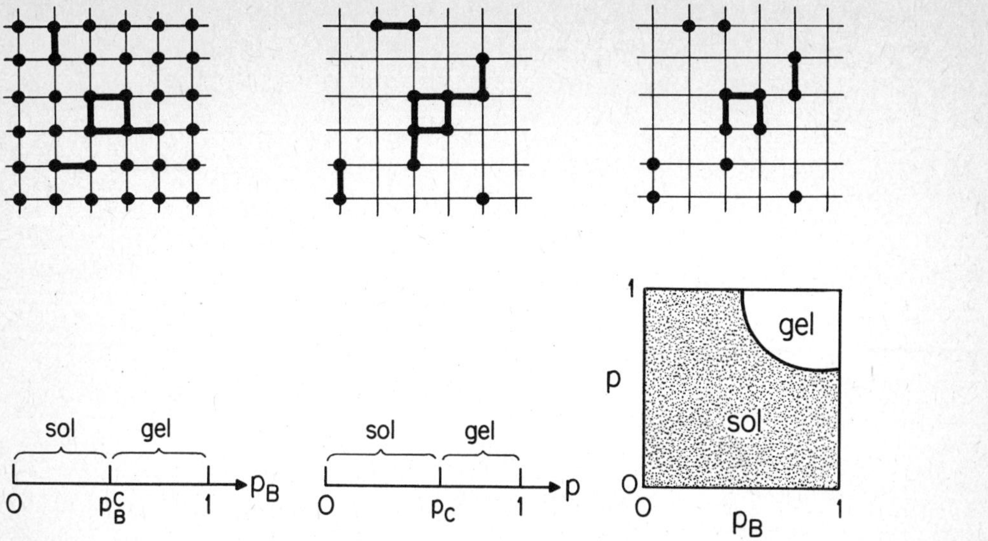

a) Bond percolation   b) Site percolation   c) Site-bond percolation

Fig.7.4a-c. Schematic illustration of (a) bond percolation, (b) site percolation, and (c) "site-bond" percolation, and their associated phase diagrams. The original Flory theory of polyfunctional condensation is an example of bond percolation, while many models of random magnets involve site percolation. Site-bond percolation is a hybrid model [7.31] that can be treated by position-space renormalization group methods [7.32,33]. When correlation is included among the sites, the resulting gelation model is capable of incorporating solvent effects and phase separation [7.34-39]

mation of an infinite branched network. Perhaps more significantly, the critical behavior of the system in the immediate vicinity of the gel point will be characterized by quite different exponents for $d < d_c$, where d_c denotes the upper marginal dimension. For gelation, d_c is believed to be six [7.30].

Thus let us consider a simple lattice model which describes polymers of arbitrary configuration, including polymers with loops. Suppose that all the sites of an infinite lattice of coordination number $z = f$ are occupied with monomers. Further, imagine that neighboring pairs of monomers can randomly form bonds, with probability p_B (Fig.7.4a). If p_B is small, the monomers group into small oligomers or "clusters". As p_B increases, the clusters grow larger; they also become more "ramified" or stringy ([7.30,40,41a]; the "ramified" incipient infinite character has been quantified in [7.41b,c]). As long as p_B is not too large, the clusters remain finite: e.g., if we pull on clusters at both "ends" of the lattice, there is no elastic restoring force. However, when p_B reaches a critical threshold, the bond network contains an infinite cluster which gives rise to a nonzero elastic modulus.

We have just described bond percolation, which may be thought of as a transcription of the Flory model of gelation from a Cayley tree pseudolattice ("no loops") to a real lattice (where loops are allowed). The analogous model of site percolation is described as follows. Imagine that the sites of the same lattice are now *randomly* occupied with probability p by monomers, and that *all* possible bonds form between nearest-neighbor monomers.[2] One now considers the clusters of sites formed in this case (Fig.7.4b). Site percolation has also proved useful in describing a variety of phenomena occurring in other systems, e.g., dilute magnets.

The position-space renormalization group methods we develop here are equally applicable to both bond and site percolation. Here we shall mainly concern ourselves with site percolation, treating bond percolation as an "extension" where appropriate.

A hybrid model, site-bond percolation (Fig.7.4c), has been proposed and successfully applied to the description of gelation in a solvent ([7.31,34,35] and references therein). In site-bond percolation, sites are occupied by monomers with probability p and by solvent molecules with probability $q = 1 - p$. Bonds connecting neighboring pairs of monomers are intact with probability p_B, which in general is dependent upon external parameters such as temperature and pH. When correlation is permitted among the monomers and solvent molecules, the resulting "site-bond correlated percolation" model describes phase separation data obtained in the gelatin-ethanol-water system [7.36].

The models described above display singular behavior in their connectivity properties at their percolation thresholds. These phenomena are true phase transitions, and renormalization group methods have proved useful in their qualitative and even quantitative description. Of course, phase transitions can only occur in infinite systems. In a finite lattice with N sites, connectivity properties become "rounded" as indicated in Fig.7.3b. For example, let us consider the probability R(p) that a network has a nonzero elastic modulus. This will happen when a single cluster (polymer) spans the entire lattice. For N small, one can enumerate all configurations of the system, and thereby calculate this probability. Thus

$$R(p) = \sum_{\text{config}} P_{\text{config}} f_{\text{config}} \quad , \tag{7.4}$$

where the summation is over all 2^N configurations of the system. Here $P_{\text{config}} = p^L q^{N-L}$ is the probability of a particular configuration of L occupied sites, while f_{config} is unity if that configuration "spans" and zero otherwise [7.42]. Our reason for focusing on R(p) is that it will serve as a recursion relation in the renormalization group approach, while f_{config} will be a "weight function". The connectivity implied by the existence of a spanning path is the physical criterion we will use to rescale probabilities.

---

2 Originally, site percolation was thought of as being a model in which all nearest-neighbor bonds are occupied, independent of the site occupancy. Of course, the connectivity properties of both models are the same.

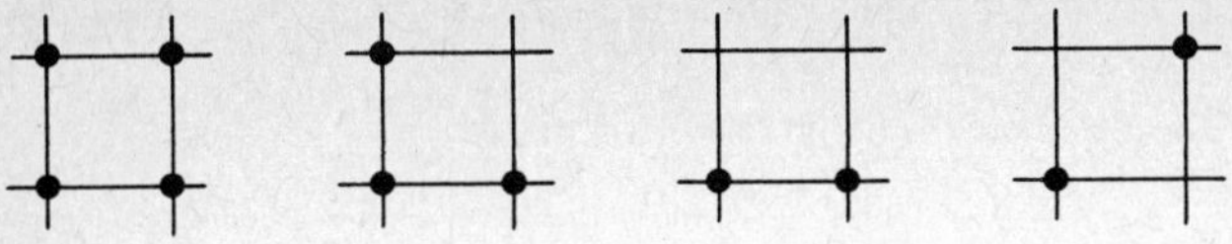

Fig.7.5a-d. Some of the 2^4 configurations that arise in the position-space renormalization group for site percolation on the square lattice using a b = 2 cell. Configurations (a-c) span from East to West, while configuration (d) does not

As an example of R(p), consider a finite lattice with only 4 sites. There are 2^4 configurations, some of which are shown in Fig.7.5. The reader can readily verify that if "spanning" is defined from East to West, then

$$R(p) = p^4 + 4p^3q + 2p^2q^2 \quad . \tag{7.5}$$

The function R(p) has been evaluated exactly for $N \leq 16$, and by Monte Carlo methods for N up to 250,000 [7.43,44]. As N increases, R(p) becomes sharper, and approaches the step function sketched in Fig.7.3 as $N \to \infty$.

7.2.2 Self-Avoiding Walks

Consider again the infinite square lattice, and choose a site at random to be called the origin. Consider now the set of all L-step random walks emanating from the origin, subject only to the global constraint that each walk not intersect itself. Because of the global constraint, such a random walk is in fact not at all random; it is customarily called a "self-avoiding walk," and given the acronym SAW [7.45,46].

We can ask many nontrivial questions about the statistical properties of SAWs. For example, let us define c_L to be the number of different self-avoiding walks of L links ("steps") that emanate from the origin; an L = 5 example is shown in Fig. 7.1a. Clearly $c_1 = 4$, since there are four different one-step walks, depending on whether the compass direction of the first step is N,E,S, or W. Similarly $c_2 = 12$ since each of the one-step walks can be continued in three different ways.[3]

Although enumerating the first few terms is easy, the enumeration of the coefficients c_L for arbitrary L is a combinatorial problem of classic difficulty. It is, however, of particular interest because the generating function, defined by

$$G(K) = 1 + \sum_{L=1}^{\infty} c_L K^L \quad , \tag{7.6}$$

3 For the square lattice the coefficients c_L have been enumerated by machine computation to order L = 24 by SYKES and collaborators [7.47].

displays a power-law singularity of the form

$$G(K) \sim |K - K_c|^{-\gamma} \quad . \qquad (7.7)$$

We call the exponent γ, because in the $n \to 0$ limit of the n-vector model, G(K) is proportional to the susceptibility above the critical temperature [7.1,2,7]. The critical parameter K_c is usually denoted $1/\mu$, where $\mu = \lim_{L\to\infty} c_L/c_{L-1}$ is called the connective constant [7.47] because asymptotically, μ monomers can connect to an existing walk to form the next larger walk. For a random walk, $\mu = z$; for a SAW, $\mu \leq z - 1$.

7.2.3 Lattice Animals

In the models we have been considering, polymers can be thought of as clusters on a lattice. Such clusters are often called *lattice animals*, because they represent all the possible shapes that can be formed out of the constituent elements. Linear lattice animals that do not self-intersect (i.e., are loopless) are just the SAWs we discussed above. However, general lattice animals may branch and may form loops (Fig.7.6). Lattice animals having the same number of elements (sites or bonds) may be further distinguished by the number of perimeter sites and/or bonds.

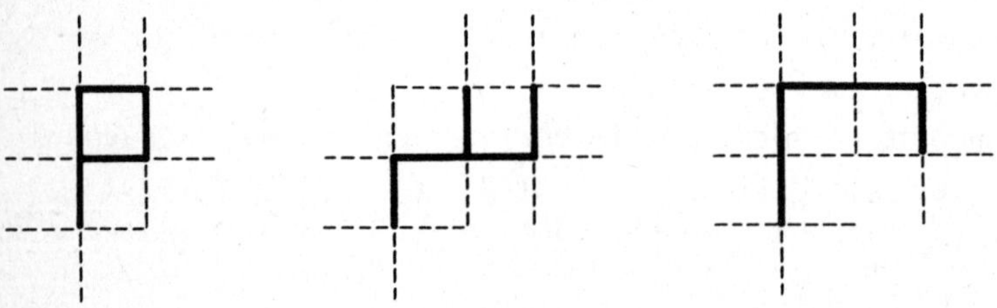

(a) (b) (c)

Fig.7.6a-c. A 5-bond branched lattice animal with looping is shown in (a), while (b) shows a 5-bond branched lattice animal without a loop. (c) shows a linear 5-bond lattice animal with neither branching nor loops, i.e., a SAW [the crossover between the lattice animal problem and the SAW problem is discussed in Sect.7.3.3]. In the *general* lattice animal problem, each of these lattice animals is weighted by a factor K^5q^t, where t is the number of perimeter bonds [cf. (7.33)]. The "pure" random animal problem obtains in the limit $q = 1$, for which all three configurations shown here are equally probable. If $q = 1 - K$, one obtains bond percolation [the crossover between pure random animals and bond percolation is also discussed in Sect.7.3.3]

In studying the structure of lattice animals, a particularly useful quantity is the probability P(L,p) that a randomly selected site or bond in the lattice is a member of an L-element cluster. In general, the functions P(L,p) are determined "recursively" from P(L - 1,p) by adding an element in all possible ways to each animal of size L - 1. For example, for *site* animals on the square lattice, the reader can easily verify that[4] $P(1,p) = pq^4$, $P(2,p) = 4p^2q^6$, and $P(3,p) = 3p^3(4q^7 + 2q^8)$.

The general form of P(L,p) is [7.30,52]

$$P(L,p) = Lp^L D(L,q) \quad , \tag{7.8a}$$

where the factor p^L arises from the L occupied sites (or bonds), and the "perimeter polynomial" $D(L,q) = \sum_t g_{Lt}q^t$ arises from the unoccupied "perimeter" that bounds the cluster. Here g_{Lt} is the number of lattice animals with perimeter t. Since D(L,q) gives the number of distinct lattice animals of L elements, grouped according to the number of perimeter elements,

$$A_L \equiv D(L,q = 1) \tag{7.8b}$$

gives the total number of L-element lattice animals.

A generating function for lattice animals, analogous to that for SAWs, is given by

$$G(K) = 1 + \sum_{L\geq 1} A_L K^L \quad . \tag{7.9}$$

Note that (7.9) subsumes (7.6), for if we restrict ourselves to linear bond animals, the generating functions are the same.

Both the site and bond animal generating functions display a power-law singularity of the form[5]

$$G(K) \sim |K - K_c|^{\theta-1} \quad . \tag{7.10}$$

The critical parameter K_c is usually denoted $1/\lambda$, where λ is called the "growth parameter" [7.48] because asymptotically $A_L/A_{L-1} \to \lambda$. Thus in an enumeration of the number of animals of size L, asymptotically one finds λ times as many animals as at the previous order (for a Cayley tree, $\lambda = z$).

---

4 The functions P(L,p) for site animals have been evaluated exactly on the square lattice for $L \leq 19$ [7.48], and on the general-d hypercubic lattice for $L \leq 8$ [7.49]. For $2 \leq d \leq 7$, P(L,p) has been evaluated for certain values of p by Monte Carlo methods for L up to about 10^6 [7.50,51].

5 Exact results exist for a one-dimensional lattice [7.53], for which $G(K)=(1-K)^{-1}$. Thus $\lambda = 1$ and $\theta - 1 = -1$. The A_L can also be evaluated exactly for the Bethe lattice [7.25], giving the result $\theta - 1 = 3/2$. It is believed [7.8-12,49] that θ takes on the value 5/2 for all systems with $d > 8$. For $1 < d < 8$, the A_L have been evaluated exactly for small values of L [7.48,49,54,55]; one finds that θ varies smoothly with d. PARISI and SOURLAS [7.10] have recently argued that $\theta(d) - 1 = \sigma(d') + 1$. Here, $\sigma(d') + 1$ is the exponent characterizing the singularity in the complex H plane [7.56] of the Gibbs potential $G(H,T_0)$ for an Ising model of dimensionality $d' = d - 2$ at a fixed temperature $T_0 > T_c$.

7.3 Position-Space Renormalization Group Treatment

In order to apply position-space renormalization group approaches to polymer models, one must first establish a connection between the sorts of "geometric phase transitions" exhibited by polymer systems and the more conventional kinds of phase transitions familiar from extensive studies of thermal critical phenomena. To this end, an important development was the observation of de GENNES [7.5] that the limit in which the number of monomers in a linear polymer tends to infinity, corresponds to the temperature of a magnetic system approaching its critical value T_c.

Polymer systems also resemble thermal systems in that both have the characteristic of self-similarity near their respective critical points. As one approaches the critical point of a magnet, fluctuations occur on larger and larger length scales until at the critical point there is no longer a "maximum" characteristic length. Viewed from a large length scale, the system is self-similar, or scale invariant. Similarly, for a polymer system at the "critical fugacity", polymers of all sizes occur, and this system is also scale invariant. At this point the polymeric system undergoes a phase transition from a state of *local* connectedness to one of *global* connectedness.

7.3.1 Percolation

a) *Basic Approach*

For the sake of clarity, we shall first discuss the position-space renormalization group treatment of percolation. It is widely appreciated that the $Q \rightarrow 1$ limit of the Q-state Potts model [7.57] corresponds to pure bond percolation [7.58]. This fact provided the initial bridge between conventional "thermal" critical phenomena and the sort of "geometrical" critical phenomena exemplified by percolation. In particular, the Potts Hamiltonian has been treated by both momentum-space renormalization group [7.17-19,59] and by position-space renormalization group [7.17,60,61], yielding predictions for percolation quantities in the limit $Q \rightarrow 1$.

A Hamiltonian, however, is not necessary for studying percolation. Indeed, in the "pure" percolation problem, the elements (sites or bonds on a lattice) are placed entirely at random with no interactions between them. However, singularities nevertheless occur in the geometric properties of the physical clusters.

In fact, the mere existence of a diverging length scale is sufficient to motivate the use of the renormalization group. STINCHCOMBE and collaborators [7.62-64] showed how a decimation procedure applied directly on the bond occupation probabilities – in which a set of vertices is summed over, leading to renormalized bond probabilities on a rescaled lattice – is equivalent to summing out degrees of freedom in the Potts partition function. Another approach for the bond problem, also treating the probabilities directly, was followed by KIRKPATRICK [7.65], who used the MIGDAL-KADANOFF approach [7.66].

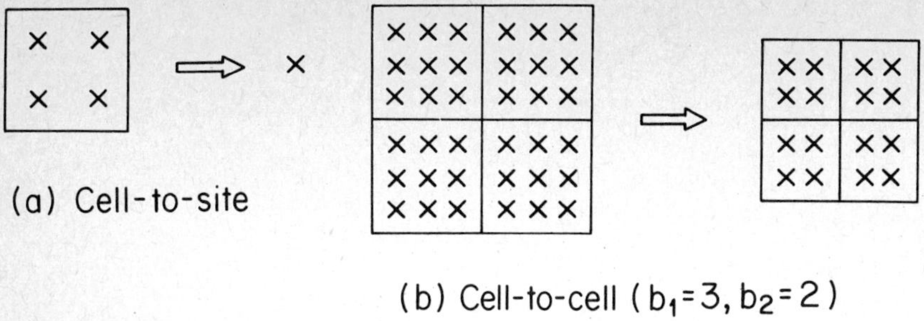

Fig.7.7a,b. Comparison between (a) the conventional "cell-to-site" construction with b=2, and (b) a "cell-to-cell" construction with $b_1=3$, $b_2=2$. By covering the lattice in two different ways, first with cells of side b_1 and then with cells of side b_2, we obtain a cell-to-cell transformation with rescaling length b_1/b_2

The numerical predictions based on the early renormalization group work were less accurate than those obtained from extrapolation of series expansions. One reason for this initial inaccuracy is related to the fact that the approaches all basically involved uncontrolled approximations.

As a simple - but readily improvable - position-space renormalization group, let us now consider a cell approach [7.42]. Partition a square lattice into cells of edge b, each containing b^2 sites; an example for b = 2 is shown in Fig.7.7a. After a renormalization transformation these cells play the role of "renormalized" sites. If the sites are occupied with probability p, then the cells may be defined to be occupied with probability p', where

$$p' = R(p) \quad . \tag{7.11}$$

The function R(p) was defined in (7.4) for an arbitrary cell and illustrated in (7.5) for the case b = 2. It is simply the probability of there being a path that spans the cell from East to West. If such a spanning path exists, then, according to this choice of weight function, the cell is occupied. Equation (7.11) serves as a simple, albeit highly approximate, position-space renormalization group transformation. The fixed points p^* of the transformation (7.11), which satisfy the relation

$$p^* = R(p^*) \quad , \tag{7.12}$$

include two trivial fixed points at $p^* = 0, 1$ and also a critical fixed point at a value that depends upon b. In this one-parameter position-space renormalization group, the critical fixed point gives an estimate for the value of the nonuniversal quantity p_c. For the b = 2 example of (7.5), $p^* = 0.62$, which should be compared with more accurate estimates, $p_c = 0.59$ ([7.67] and references therein).

To calculate the critical exponent ν_p from this simple transformation, we note that all lengths in the rescaled system have been reduced by a factor of b from the lengths in the original system.[6] Hence the connectedness length ξ transforms as

Footnote 6 see next page

$\xi' = b^{-1}\xi$. Since $\xi \sim |p - p_c|^{-\nu_p}$ and p_c corresponds to p^*, we obtain

$$|p' - p^*|^{-\nu_p} = b^{-1}|p - p^*|^{-\nu_p} \quad . \tag{7.13}$$

On the other hand, we can subtract (7.12) from (7.11). Because the transformation R(p) is analytic near p^*,

$$p' - p^* = R(p) - R(p^*) = \lambda(p - p^*) \quad , \tag{7.14}$$

where $\lambda = dR/dp|_{p^*}$. If (7.14) and (7.13) are to be consistent, we must have $\lambda^{-\nu_p} = b^{-1}$ or

$$\nu_p = \ln b/\ln\lambda \quad . \tag{7.15}$$

For the b = 2 example in question, we find $\nu_p \simeq 1.6$, which should be compared with the possibly exact value $\nu_p = 4/3$ [7.68-70].

This completes our simple position-space renormalization group example for the case b = 2. The results for p_c and ν_p are not expected to be very accurate, due to the "one-shot" nature and the simplicity of the approximation. It is, however, useful to try to understand the source of this error; in so doing, a natural method for compensating for the error will arise. Figure 7.8 shows a possible occupancy configuration for four adjacent cells. Notice that cells B and C are connected in the original problem, but not in the rescaled problem. Conversely, cells A and D are not connected in the original problem, but are connected in the rescaled picture. Thus with the simple connectivity rule, the rescaled cell problem may not faithfully mirror the connectivity of the original problem.

In the conventional renormalization group approach, intercell connections can be "patched up" by introducing additional couplings (further neighbor, multisite, etc.). The entire set of couplings must then be renormalized. This leads to the usual ideas about relevant and irrelevant operators, and flow trajectories. This method is most useful when there is a crossover effect to study (see Sect.7.3.3). In pure percolation, all the additional couplings are irrelevant, and their main effect is to complicate the calculation. The number of parameters grows quickly,

6 The connectedness length ξ_p characterizes the exponential decay with r (near but not at p_c) of the pair connectedness function C(r), which in turn gives the probability that a site at position r is occupied and connected to a site at the origin. The singularity in ξ_p at the "percolation critical point" ($p=p_c$, $L=\infty$) and its behavior near the critical point is described by the exponent ν_p if the variable $p-p_c$ is used, and by the exponent ν if the variable L is used. The exponents ν_p and ν are related by $\nu_p = \nu\Delta$, where Δ is the "gap exponent," and $\Delta = \beta\delta = \beta + \gamma$ by scaling. For largely historical reasons, the exponent ν is denoted ρ in much of the percolation literature, while the exponent ν_p is denoted ν.

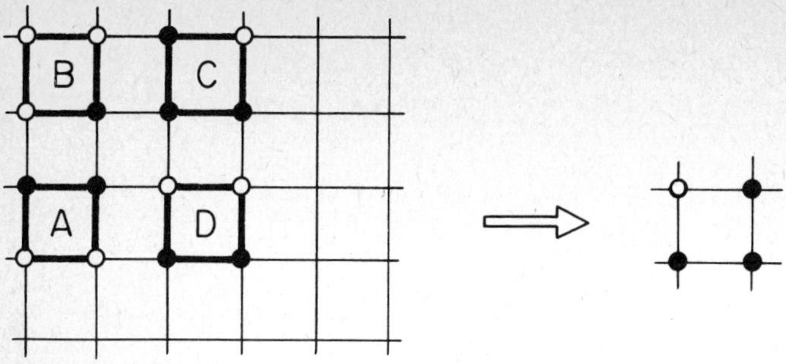

Fig.7.8. Rescaling a lattice by forming cells out of groups of sites. Four-site cells on the square lattice are shown, with dots representing occupied sites. In this example, cells C and D are both "occupied" (each cell can be traversed), and therefore connected on the *cell* level. On the *site* level, however, we cannot connect them. On the other hand, cells A,B, and C are connected on the *site* level, though only the next-neighbor cells, A and C, are occupied. These incorrect connections force one to either introduce nearest-neighbor, further-neighbor, and multisite probabilities, or to go to larger cells where the "interfacing" error plays a smaller relative role [7.42]

but there is no guarantee that there will be a corresponding return in increased accuracy. To bypass these complications, we next consider a different approach in which only one parameter, p, is kept.

b) *Extensions*

1) Large-Cell Monte Carlo Renormalization Group. In order to improve upon the one-parameter, "small-cell" method, we note that the error introduced by incorrect intercell connections is essentially a surface effect; as such, it should get smaller as the cell size gets bigger [7.71]. This in fact happens. Thus our approach here will be to retain the single-parameter position-space renormalization group, but increase b systematically. For b = 3, the summation in (7.4) extends over $2^9 = 512$ configurations, while for b = 4 and 5, there are 2^{16} and 2^{25} configurations respectively. Although 2^{25} is a rather large number, one may use a computer to determine f_{config} for each of these configurations, and thereby obtain R(p) exactly. The values of $p^*(b)$ and $\nu_p(b)$ become increasingly accurate with larger b, but the convergence in b is rather slow. Nevertheless, we can estimate the critical parameters fairly accurately, based on the largest values of b (b = 5) for which R(p) may be computed in closed form.

However, we can do better still! The systematic improvement of the small cell results with b suggests that we *extrapolate* our finite-b results to $b \to \infty$, somewhat in the spirit of extrapolations based on exact enumeration methods. The renormalization group extrapolation procedure parallels the phenomenological theory of finite-size scaling([7.72]; see also the recent justification of finite-size scaling in [7.73]). In this sense, the errors in these extrapolated estimates are less "uncontrolled" than those made on the basis of other renormalization group methods of successive approximations — such as higher order cumulant expansions — for which there is no systematic procedure for extrapolating successive estimates for the critical parameters.

For example, to extrapolate the successive estimates of $p_c(b)$, finite-size scaling considerations [7.43,44] suggest

$$|p^*(b) - p_c| \sim b^{-1/\nu_p} \quad . \tag{7.16a}$$

Furthermore, from considerations of the error in the position-space renormalization group eigenvalue, a likely form for $\nu_p(b)$ is [7.43,44,74]

$$|\nu_p(b) - \nu_p| \sim A_1(\ln b)^{-1} + A_2(\ln b)^{-2} \quad . \tag{7.16b}$$

Thus the error will decrease as b increases in a *predictable* fashion, and a sequence of values of $p^*(b)$ and $\nu_p(b)$ can be used to extrapolate the limiting $(b \to \infty)$ behavior.

However, for accuracy comparable to series, one must extrapolate using values of $p^*(b)$ and $\nu_p(b)$ for $b > 5$. To do this, a large-cell Monte Carlo renormalization group was developed [7.43,44]. This method shows the close connection of one-parameter position-space renormalization groups with finite-size scaling. The large-cell calculations and extrapolations for site percolation have been carried out for the square lattice [7.43,44], and for the triangular lattice [7.74], using an algorithm based on the work of HOSHEN and KOPELMAN [7.75]. This algorithm tests which realizations span, and hence have $f_{config} = 1$. One finds [7.44] that R(p) can be calculated with remarkable accuracy by sampling comparatively few of the 2^N $(N = b^2)$ of the possible configurations.[7]

The results of the square lattice calculations are summarized in Fig.7.9a [which is based on (7.16a)], and in Fig.7.9b [which is based on (7.16b)]. Actually, two sets of curves are shown in Fig.7.9a. The set labelled R_1 corresponds to the connectivity weight function described earlier, which assigns the value $f_{config} = 1$ if the configuration spans the cell in the East-West direction. The set labelled R_0 differs in that $f_{config} = 1$ if the configuration spans *either* in the East-West *or* in the North-South direction. Notice that even for the largest cell size considered, the estimates $p^*(b)$ and $\nu_p(b)$ — though accurate by usual position-space renormalization group standards — are still not as accurate as one would desire; in fact, they are only beginning to approach the accuracy available from series expansions [7.30,52]. However, since the dependence of $p^*(b)$ and $\nu_p(b)$ follows

---

7 For example, for b = 500, N = 250,000; there are roughly $10^{75,000}$ configurations! The function R(p) was calculated by averaging the results of only 10^3 total realizations. The sharpness of R(p) for a system with this many sites made it highly probable that when R(p) was calculated again, using 10^3 *different* realizations, the two calculations agreed to remarkable accuracy.

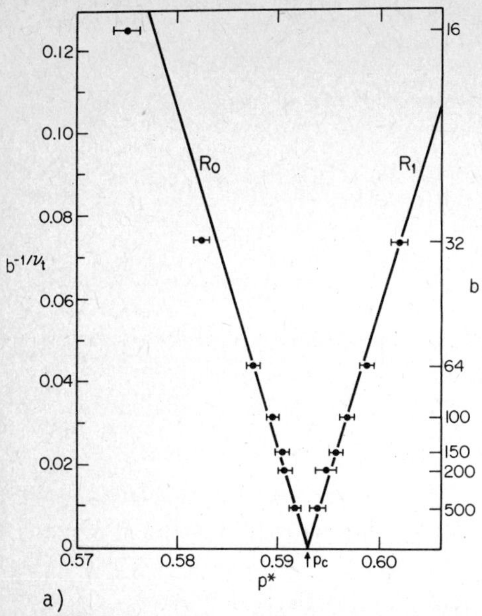

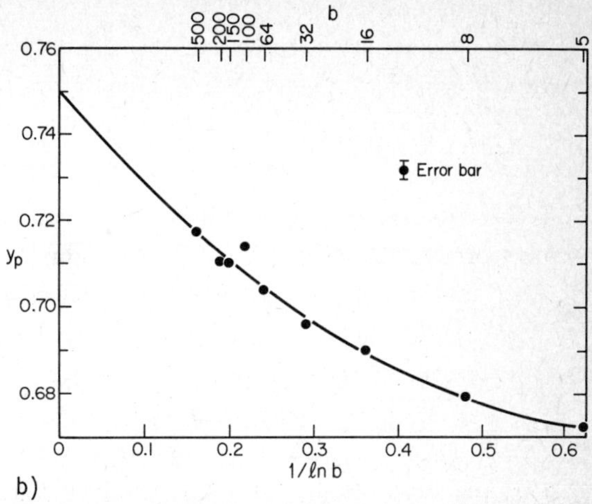

Fig.7.9. (a) Extrapolation of the sequence of fixed points $p^*(b)$ obtained from connectivity rules R_0 and R_1 using (7.16a). We have chosen the trial value of $\nu_t = 4/3$, for this plot. Both curves approach approximately the same final estimate of p_c as $b \to \infty$. (b) Quadratic extrapolation against $1/\ln b$ for y_p, using rule R_0 and (7.16b). The error bar shown is representative. The solid curve is the least-squares fit, $y_p(b) = 0.7490 - 0.0221/(\ln b) + 0.0157/(\ln b)^2$

(7.16), the finite-b results can be extrapolated to the limit $b \to \infty$. We see from Figs.7.9a,b that the $b \to \infty$ extrapolation provides estimates for p^* and ν_p that differ substantially from any of the finite-b values. These *extrapolated* estimates are as accurate as series estimates, and are possibly even somewhat more accurate. One obtains $p_c = 0.5931 \pm 0.006$ and $\nu_p = 1.33 \pm 0.01$, the latter result requiring that we keep quadratic terms in (7.16b)[8]. Moreover, the fact that the estimates derived from "connectivity rule R_0" agree with those obtained from "connectivity rule R_1" adds further confidence to the results obtained. The result for ν_p agrees well with the widely accepted den NIJS conjecture [7.68].

2) Ghost Field. The above procedure gives the scaling power $y_p = \nu_p^{-1}$ analogous to the thermal scaling power in magnetic critical phenomena. To obtain all the percolation exponents, one must also calculate the scaling power y_h, which is analogous to the magnetic-field scaling power.

In percolation, the role of a magnetic field may be played by a ghost field ([7.53] and references therein). This can be defined in a variety of fashions

8 Quadratic terms were omitted in [7.44], and linear extrapolation resulted in the higher estimate $\nu_p = 1.354 \pm 0.015$. Quadratic terms are included in [7.74] and in Fig.7.9b, thereby reducing the estimate for ν_p.

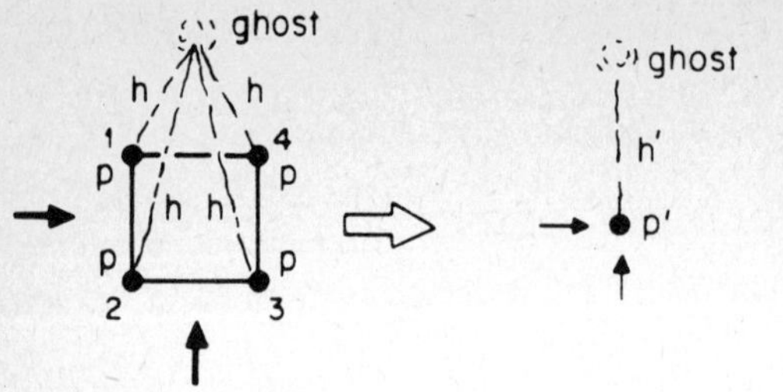

Fig.7.10. Illustration of the rescaling of a b=2 cell for site percolation on the square lattice in the presence of a nonzero "ghost field" h. The four original sites, each occupied with probability p, are transformed into a single renormalized site occupied with probability p'. The ghost bonds, which join to every site, are transformed into a bond with probability h' [7.43]

(see, e.g., the discussion in [7.52]). The simplest, conceptually, is to imagine that there is an extra "ghost site", exterior to the lattice, which is connected to every site in the lattice by a "ghost bond" which is occupied with probability h.

Equation (7.4) still determines R(p), except now f_{config} depends upon h. For example, consider the simple b = 2 case. For the configurations shown in Figs. 7.5a-c, $f_{config} = 1$ regardless of h. However, for the configuration shown in Fig. 7.5d, $f_{config} = h^2$, since if the two ghost bonds connected to the two occupied sites are both intact, then the configuration spans through the ghost site. Thus

$$p' = R(p,h) = R(p,h = 0) + 2p^2q^2h^2 \quad , \tag{7.17}$$

where R(p,h = 0) is given by (7.5).

It is also necessary to obtain a recursion relation for the ghost field h'. To determine this, notice that the presence of the ghost bonds allows the possibility of connecting to the ghost. This suggests that we can renormalize h by calculating the rescaled probability of being connected to the ghost. This connectivity is most appropriately expressed in terms of the probability of entering the cell *and* reaching the ghost, given by p'h' (Fig.7.10). This can be written in an obvious generalization of (7.4) as

$$p'h' = \sum_{config} P_{config} f_{config} \quad , \tag{7.18a}$$

where now the summation is over all 2^{2b^2} configurations of the b^2 sites and b^2 ghost bonds. For the case b = 2 shown in Fig.7.10, one finds [7.44]

$$p'h' = p^4Q_4 + 4p^3qQ_3 + p^2q^2(h + 5Q_2) + 3pq^3h \quad , \tag{7.18b}$$

where $Q_j \equiv 1 - (1 - h)^j$ is the probability that at least one of the j ghost bonds connecting the ghost to a set of j sites is intact. The recursion relation (7.18b) is derived by enumerating all connected paths from 2 adjacent edges of the cell to the ghost and calculating the sum of the probabilities for each such path (see Figs.7.5,10).[9] (Footnote 9 see next page).

By implementing this general procedure, and by using large-cell Monte Carlo methods for cells of size up to N = 250,000 sites, one can make a reliable extrapolation for y_h [7.44]. The result,

$$y_h = 1.898 \pm 0.003 \quad , \tag{7.19a}$$

is in remarkably good agreement with the closed-form extended den Nijs conjecture [7.76,77],

$$y_h = 91/48 = 1.896 \quad . \tag{7.19b}$$

3) Cell-to-Cell Transformation. Until now, we have considered the conventional position-space renormalization group approach of transforming from a system of cells to a new system of sites. This "cell-to-site" transformation is illustrated schematically in Fig.7.7a. However, one can also consider a transformation in which one passes from a system of cells of edge b_1 to a system of cells of edge b_2 [7.43] (Fig.7.7b). Such an implicit "cell-to-cell" transformation enables one to have a rescaling length b_1/b_2 which may be made close to unity.

We will illustrate the cell-to-cell transformation for the case of zero ghost field. There is then only one parameter, the occupation probability p. To calculate the critical exponent ν_p, we define $p_j = R(b_j;p)$ to be the renormalized occupation probability for a cell of size b_j $(j = 1,2)$. Then the connectedness lengths of the two systems are related by

$$\xi(p_2) = (b_1/b_2)^{-1}\xi(p_1) \quad . \tag{7.20}$$

Hence the equation analogous to (7.13) is

$$|p_2 - p^*|^{-\nu_p} = (b_1/b_2)^{-1}\,|p_1 - p^*|^{-\nu_p} \quad , \tag{7.21}$$

where p^* is the fixed point of the cell-to-cell transformation. Note that (7.21) reduces to (7.13) if $b_2 = 1$ (and hence $p_1 = p$). The analog of (7.14) is

$$p_j - p^* = R(b_j;p) - R(b_j;p^*) = \lambda_j[p - p^*] \quad , \tag{7.22}$$

where $\lambda_j = [dR(b_j;p)/dp]_{p^*}$. Substituting (7.22) into (7.21), one obtains $(\lambda_2/\lambda_1)^{-\nu_p} = (b_1/b_2)^{-1}$, or

$$\nu_p = \ln(b_1/b_2)/\ln(\lambda_1/\lambda_2) \quad . \tag{7.23}$$

---

9 The first term is the product of the probability that all 4 sites are occupied, and that at least *one* of the 4 bonds connecting these four sites to the ghost site is intact. The second term is the probability that any 3 of the sites are occupied, and that at least one of the 3 bonds connecting these 3 occupied sites to the ghost site is intact. The third term is more complicated, since not all 6 of the configurations of two occupied sites are equivalent. If sites 2 and 4 in Fig.7.10 are occupied (as in Fig.7.5d), then the probability of being able to "get to the ghost" is simply p^2q^2h, since the ghost bond on site 2 must be intact, while the state of the ghost bond on site 4 is irrelevant. For the remaining 5 configurations of two occupied sites, the probability is $p^2q^2Q_2$, since either of the two ghost bonds must be intact. Finally, of the four configurations with only one occupied site, three of these will connect to the ghost, provided the corresponding ghost bond is intact. Each of these occurs with probability pq^3h.

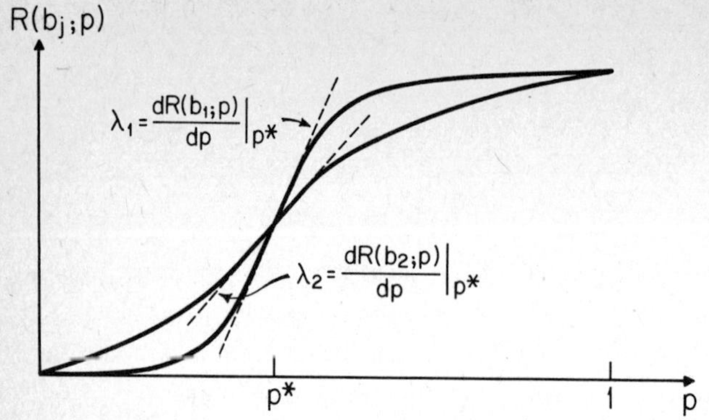

Fig.7.11. The dependence on p of the two renormalization group transformations $p_1=R(b_1;p)$ and $p_2=R(b_2;p)$. The fixed point of the transformation from p_1 to p_2 occurs where the two curves intersect. The exponent ν_p is obtained from the ratios of the slopes of these functions at the point of intersection using (7.23)

Again, (7.23) reduces to (7.15) if $b_2 = 1$ since then $\lambda_2 = 1$.

The location of the fixed point p^* and the calculation of the exponent ν_p is indicated geometrically in Fig.7.11. The cell-to-cell transformation has been used [7.43] to calculate $p^*(b_1,b_2)$ and $\nu(b_1,b_2)$ for all values of b_1,b_2 such that $b_2 < b_1 \leq 5$. It is found that the estimates are considerably better than for the corresponding cell-to-site transformations. Thus the strength of the cell-to-cell method is that it provides higher accuracy for a given amount of calculation—on the order of a cell-to-site calculation with b about 100. This higher accuracy probably results from the cancellation of similar errors in the functions $R(b_1,p)$ and $R(b_2,p)$. Additionally, extrapolation of the cell-to-cell scaling powers appears to converge as $1/b$ rather than $1/\ln b$.

For $b_1 > 5$, the transformations $R(b_j;p)$ are not known exactly. The combined approximations obtained from Monte Carlo simulations of two cell-to-site transformations give considerably larger error bars for the cell-to-cell transformation than for the corresponding cell-to-site transformation. This is apparent from Fig. 7.11 and the form of (7.23). A cell-to-cell Monte Carlo renormalization group that avoids this problem is also possible, but at present the cell-to-cell method is most useful when one has closed-form expressions for $R(b_j;p)$. For a discussion of a cell-to-cell approach in the context of thermal phase transitions, see Chap.3

4) Bond Percolation. For bond percolation, it is not immediately obvious how to choose a cell that covers the lattice with bonds, and rescales to another bond. One very useful choice is the family of cells [7.42] illustrated schematically for b = 2 in Fig.7.12. A simple weight function can be defined such that if the cell can be traversed horizontally, then the renormalized horizontal bond is considered to be intact, while if the cell can be traversed vertically, then the renormalized vertical bond is intact. The resulting recursion relation for b = 2, which is analogous to (7.5), is then

$$R(p) = p^5 + 5p^4q + 8p^3q^2 + 2p^2q^3 \quad . \tag{7.24}$$

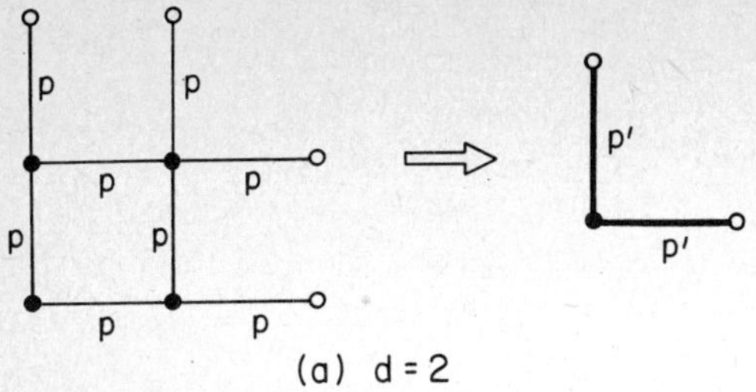

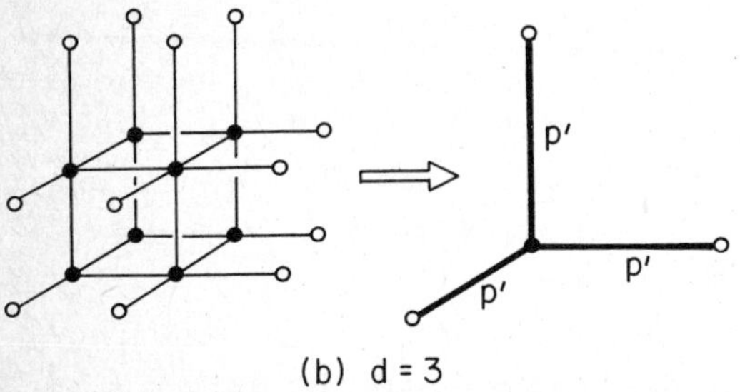

Fig.7.12a,b. The simplest b=2 cells for bond percolation for (a) the square lattice, and (b) the simple cubic lattice. The rescaled bonds are indicated by heavy lines and occur with probability p'

The nontrivial fixed point of the transformation $p' = R(p)$ is at $p^* = 0.5$, which agrees with the exact result, $p_c = 1/2$, for bond percolation on the square lattice. In fact, BERNASCONI [7.78] showed that this family of cells has the same self-duality as the square lattice itself, and thus $p^* = 1/2$ for all b. Unfortunately, the self-duality does not guarantee anything for the exponent. The above $b = 2$ recursion relation results in the reasonable approximation $\nu_p = 1.43$. KIRKPATRICK [7.79] has performed large-cell calculations for R(p) by Monte Carlo methods, with $b = 16$, 64, 200, and 512. However, he used a slightly different cell that does not preserve self-duality. Thus his sequence $p^*(b)$ is not fixed at 1/2, but only approaches 1/2 as $b \to \infty$. From his sequence $\nu_p(b)$, he estimates $\nu_p = 1.365 \pm 0.015$, which is just outside the now favored result $\nu_p = 4/3$. More recently, LOBB and KARASEK [7.80] calculated the sequence $\nu_p(b)$ using the self-dual cells and using many more cell sizes, to obtain an extrapolated values $\nu_p = 1.333 \pm 0.009$. Also their result, $y_h = 1.900 \pm 0.017$ for $d = 2$ bond percolation, agrees with the result in (7.19a) for site percolation. Also, MAGALHÃES et al. [7.81] and HEERMANN and STAUFFER [7.82] used this approach on the square bond problem, with good results.

5) Three Dimensions. Bond percolation calculations on the simple cubic lattice have been carried out analytically for $b = 2$ ([7.42]; see also [7.83] and references therein), using the cell shown in Fig.7.12b. KIRKPATRICK [7.79] extended this work to large cells and extrapolated to obtain $\nu_p = 0.845 \pm 0.015$. More recent calculations give $\nu_p = 0.88 \pm 0.01$ [7.82]. These position-space renormalization group numbers may be compared with the estimate $\nu = 0.83$ from series enumeration methods [7.84].

6) *Other Percolation Properties*

We have thus far restricted our discussion to the application of position-space renormalization group methods to the derivation of the percolation threshold and the two scaling powers y_p and y_h. The methods we have been describing may also be used to obtain other percolation quantities. Thus, e.g., one can calculate the exponent that describes the fashion in which the electrical conductivity of a percolation network approaches zero as $p \to p_c$ from above; the accuracy obtained in two dimensions [7.78,85-88] is comparable to that from series expansions [7.89].

In part because of its importance in understanding the physical phenomena occurring in electrical conductivity, the "backbone" of a bond percolation cluster has received considerable attention. SHLIFER et al. [7.90] have used large-cell Monte Carlo position-space renormalization group to calculate the scaling powers y_p and y_h for the backbone; their estimates agree quite well with independent Monte Carlo computer simulations [7.91,92].

VICSEK and KERTÉSZ [7.21] used large-cell methods to describe the connectivity properties of a continuum two-dimensional random system consisting of overlapping discs; their results are in general accord with Monte Carlo and series calculations [7.22], and support the notion that continuum systems – such as polymer gels – are in the same universality class as lattice systems.

KINZEL [7.93] has used the position-space renormalization group approach to calculate the cluster-size distribution function P(L,p), which is of particular interest because it obeys a scaling relation, the "cluster-number scaling hypothesis" [7.30].

7.3.2 Self-Avoiding Walks

a) *Basic Approach*

We begin by recalling the generating function G(K) defined in (7.6) for c_L, the total number of SAWs of L steps. As discussed above, G(K) for the SAW problem is singular at $K = K_c$, thus signalling the presence of a phase transition.

To formulate a position-space renormalization group for SAWs, one may follow the same general procedure used above to construct the position-space renormalization group for bond percolation. In percolation, the connectivity transition is signalled by the formation of an "incipient" infinite cluster [7.41,92,94]. This is a "critical" object which occupies a zero fraction of an infinite lattice, but which is globally connected. The weight function in percolation is chosen to detect this global connectivity. That is, cells with connected paths of bonds spanning the cell are mapped into renormalized bonds. Similarly, for SAWs we begin by recognizing that the critical object at the SAW threshold is a walk that is infinite in length. Moreover, this walk also occupies a zero fraction of an infinite system. Evidently the incipient infinite SAW plays the same role as the incipient infinite cluster in percolation. Accordingly, we might imagine that a "connectivity rule" weight function analogous to the one used in bond percolation should also be applicable to the SAW problem. In such an approach, a renormalized bond is present if there is a connected, self-avoiding, path that spans a cell. This defines one possible weight function. The sum of the statistical weights for spanning SAWs has the form of a generating function, which is restricted due to the weight function. It is given by [7.95]

$$G(K) = \sum_{\text{config}} K^L f_{\text{config}} \quad . \tag{7.25a}$$

Here the summation is over all 2^B configurations of the B cell bonds. The function f is the weight function: f = 1 if the configuration represents a spanning SAW, and f = 0 otherwise. Different weight functions will differ in the definition of "spanning". To obtain a recursion relation, we equate this to the generating function on the cell level, where $G'(K') = K'$.

Let us consider, for example, the "corner rule" weight function [7.96-98] on the square lattice. The starting point of the SAW is fixed to be a corner of the cell, e.g., the lower-left corner, as shown in Fig.7.13 for b = 2. Walks that leave the cell by way of the right edge rescale to a single horizontal step, while walks leaving by the top edge rescale to a single vertical step. By symmetry, it suffices to consider only the 4 walks that rescale to a vertical step, and these are shown in Fig.7.13b. Weighting each walk with a factor of the monomer fugacity K for each of its bonds, one finds the recursion relation

$$K' = K^4 + 2K^3 + K^2 \quad . \tag{7.25b}$$

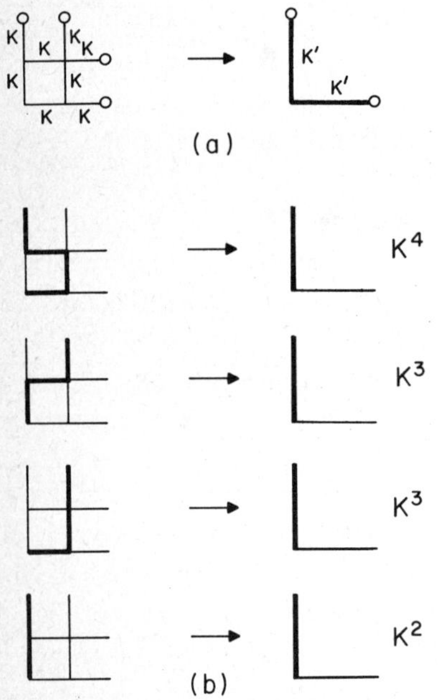

Fig.7.13. (a) The simplest b = 2 cell for the SAW problem on a square lattice (cf. Fig. 7.12a). (b) The four SAW configurations that contribute to a rescaled vertical step, which has fugacity K'

The structure of (7.25b) is analogous to the structure of the recursion relation (7.24) for bond percolation. We see that there exist 3 fixed points, two trivial ones at $K^* = 0, \infty$, and a critical one at a finite value of K, whose location is dependent on the details of the approximation. We interpret them as follows:

1) There is an "empty lattice" fixed point $K^* = 0$; the recursion relation (7.25b) faithfully rescales $K = 0$ into another empty lattice with $K' = 0$. For all $K < K^*$, SAWs are finite in length, and they shrink upon each rescaling. The limit of this process is a polymer of size zero, as described by the $K^* = 0$ fixed point.

2) There is a "full lattice" fixed point at $K^* = \infty$. In bond percolation, the full lattice has every bond intact, while in the SAW problem, every *site* of the lattice is visited, and the SAW densely packs the lattice.

3) There is a critical fixed point at $K^* = 0.466$ (for the corner rule) corresponding to the critical fugacity $K_c = 1/\mu$; the best estimate from exact enumeration procedures is $K_c = 1/\mu = 0.379003 \pm 0.000015$ [7.47]. Moreover the divergence of ξ as $K \to K^*$ is given by $\nu = \ln b/\ln\lambda = 0.715$, where λ is the slope of $K'(K)$ at K^*; this should be compared with numerical estimates for ν in two dimensions in the range 0.74-0.75 [7.1].

There are several types of approximation in this renormalization procedure. For example, there is the restriction placed on the SAW that its starting point be fixed at the corner of the cell. This is an approximation since it is unlikely that a traversing SAW in one cell will terminate at the starting point for an SAW in the next cell. Also, there exist configurations that can contribute to an infinite SAW that do not span the cell at all, e.g., a walk that leaves the cell by the same edge at which it enters. To avoid these sorts of problems, one possibility is to choose a better weight function. For example, one can in principle average over all possible starting points of a walk in the cell. However, because the starting points are not equivalent, a weight function formally akin to the transfer matrix has proven more useful. The best results seem to be obtained if one uses periodic boundary conditions, which correspond to wrapping the cell onto a cylinder or torus [7.98,99]. A toroidal cell is translationally invariant, and all starting points are equivalent. The weight function then counts all walks that wind around the torus once with respect to the origin. Thus for the simple case $b = 2$, one has [7.98]

$$K' = 2K^4 + 4K^3 + K^2 \quad . \tag{7.26}$$

The nontrivial fixed point $p^* = 0.366$ and the estimate $\nu = 0.690$ are both reasonable for such a small cell, though ν is worse than by the corner rule.

In principle, one can obtain improved accuracy by using a larger *cluster* of cells, and appropriately enlarging the parameter space to include additional fugacities for longer-range links joining further-neighbor monomers (see, e.g.,

[7.100]). These additional parameters are required to maintain the proper *inter-cell* correlations upon rescaling. This approach is somewhat cumbersome because the number of parameters quickly becomes very large. To improve the accuracy of the small-cell calculation, but still maintain the simplicity of the one-parameter approach, it is natural to consider using larger cells, as was discussed in the section on percolation.

With all our weight functions, configurations within the cell are accounted for to a good approximation, but connection of chains between cells is treated inaccurately. While it seems this should introduce an error proportional to the surface-to-volume ratio of the cell, it is not clear whether this is literally the case. Nevertheless the results of the closed-form position-space renormalization group with small cells do indicate a trend toward the "correct" asymptotic limit [7.98].

b) *Extensions*

1) Large Cells. Based on the trend in the small-cell results and on the analogy with percolation, the above approach has been extended to large cells using a "constant fugacity" Monte Carlo method in which K, the bond fugacity, is the basic parameter as opposed to the number of steps N used in conventional simulation methods [7.98]. Through this approach, chains of order 10^3 bonds can be simulated in an a priori unbiased fashion on the square lattice with only moderate use of computer time. The efficiency of this method compares well with standard methods of Monte Carlo simulation of chains [7.45,46].

By this method, the recursion relation for toroidal cells with b up to 150 were studied (Table 7.2). By extrapolating the sequence of estimates according to (7.16b), one obtains the estimate $\nu = 0.756 \pm 0.004$. This value may be compared with (7.2), the Flory approximation $\nu_F = 3/4$. It is not clear whether the present results are sufficiently reliable to safely exclude the Flory value. For example, the sequence $\nu(b)$ obtained using the first weight function we discussed – the corner rule – is not monotonic; hence it is not clear that the toroidal cell results are correct just because they *are* monotonic!

2) Three-Dimensional Systems. The above procedure can be readily extended to any dimension, with a corresponding increase in the labor of calculating the recursion relation K'(K). The corner rule has been used [7.101] to study the d = 3 simple cubic lattice for b = 2 and 3. For b = 2,

$$K'(K) = 10K^8 + 16K^7 + 14K^6 + 12K^5 + 8K^4 + 4K^3 + K^2 \quad . \tag{7.27}$$

The nontrivial fixed point is at $K^* = 0.297$, which may be compared to the most reliable estimate $K_c = 1/\mu = 0.2135 \pm 0.0001$ [7.102]. The eigenvalue obtained from (7.27) results in the prediction $\nu = 0.588$, which agrees precisely with the highly accurate field theory prediction $\nu = 0.588 \pm 0.0015$ [7.103], and with the value

Table 7.2. An example of Monte Carlo renormalization group results. The sequence shown is for the case of linear polymers (toroidal rule). Similar sequences exist for percolation [Ref.7.44, Tables 7,8]. In the large cell results, the numbers in parenthesis give the uncertainty in the last decimal place shown. Note the convergence toward accurate extrapolated results. Adapted from [7.98]

(a) cell-to-bond

b	K^*	ν
2	0.3660	0.6897
3	0.3707	0.7050
4	0.3725	0.7122
5	0.3731	0.7165
10	0.3756(3)	0.7258(1)
20	0.3768(2)	0.7318(1)
40	0.3779(1)	0.7360(3)
80	0.3784(1)	0.7400(7)
150	0.3787(1)	0.7417(5)
extrapolation	0.3791(1)	0.756(4)

(b) cell-to-cell

b'/b	K^*	ν
2/1	0.3660	0.6897
3/2	0.3747	0.7329
4/3	0.3754	0.7412
5/4	0.3759	0.7443
extrapolation	0.3791	0.755

$\nu = 0.586 \pm 0.004$ [7.104] obtained from careful re-analysis of much of the existing experimental data. Of course, this accurate agreement may only be coincidental. In fact, when one extends the calculation to $b = 3$ [7.101], one finds that although $K^* = 0.276$ is closer to the series estimate, $\nu = 0.581$ is somewhat less remarkable. Additional results using larger cells and other weight functions are clearly needed.

3) Cell-to-Cell Transformation. Following the spirit of the percolation problem, one can also derive a cell-to-cell transformation in which a cell of edge b is rescaled to a cell of edge b'. The results of such a cell-to-cell transformation, with $b'/b = 6/5$, give values comparable in accuracy to a cell-to-bond transformation with $b \sim 100$ [7.98]. This is similar to the increase in accuracy achieved in percolation using the cell-to-cell transformation. Furthermore, extrapolating the $d = 2$ cell-to-cell results versus $\ln b/b'$, yields values for ν in the vicinity of 0.755 depending on the weight function used. These numbers agree quite well with the large-cell extrapolation using the toroidal rule.

4) The $b \to 1$ Limit. Although it is not in general possible to calculate recursion relations in closed form for arbitrary b and d, progress can be made if we take the limit $b \to 1$. A similar approach for percolation was considered by SHAPIRO [7.105,106]. For example, for the corner rule, one obtains a lower bound to K' by noting that there is at least one SAW of any length $n > b$ which can span the cell. Thus [7.98]

$$K'_{\text{lower bound}} = \sum_{n=b}^{b^d} K^n < K'_{\text{exact}} \quad . \tag{7.28a}$$

Similarly, an upper bound is obtained by noting that the number of spanning n-step SAWs is bounded by an exponential. That is,

$$K'_{\text{upper bound}} = \sum_{n=b}^{b^d} b^{n-b} K^n > K'_{\text{exact}} \quad . \tag{7.28b}$$

The sums in (7.28) can both be evaluated, and as $b \to 1$, they become identical:

$$K'_{\text{inf}} = K + (b-1)(1-Kd)K(\ln K)/(1-K) \quad . \tag{7.29}$$

This transformation, of course, still contains vestiges of the approximation inherent in the finite-cell approach. From this recursion relation, one finds

$$\nu = (d-1)/(d \ln d) \quad . \tag{7.30}$$

This formula provides fairly reasonable results: ν = 1, 0.721, 0.607, and 0.541 in d = 1,2,3, and 4 respectively.

Transformations obtained in this way for *percolation* [7.65,106] are identical to the Migdal-Kadanoff renormalization-group recursion relations, which are exact in d = 1; likewise, the cell approach is exact in d = 1. Thus, for $d \to 1$ (7.30) may also be exact. To first order in $\varepsilon = d - 1$, (7.30) gives $\nu = 1 - \varepsilon/2$. In contrast, the Flory formula (7.2) gives $\nu_F = 1 - \varepsilon/3$. Thus, just as for $d = 4 - \varepsilon$ [7.5], the Flory formula appears to give too large a value for ν.

7.3.3 Lattice Animals

a) *Basic Approach*

The generating function for lattice animals (7.9) is formally like the generating function (7.6) for SAWs, except that A_L now includes *all* configurations of connected clusters. Hence to formulate a position-space renormalization group for lattice animals, one need only follow the arguments of the previous section, replacing SAWs by lattice animals in the configurations we sum over. For the square lattice with b = 2, one finds for bond animals [7.97]

$$K' = K^8 + 8K^7 + 21K^6 + 24K^5 + 14K^4 + 4K^3 + K^2 \quad . \tag{7.31}$$

The fixed point of this transformation is at K = 0.270, which may be compared with the result $1/\mu = 0.192 \pm 0.002$ [7.48]. The corresponding exponent characterizing the singularity in the mean "animal" radius is $\nu = 0.571$, which is smaller than the values 0.61 obtained by PARISI and SOURLAS [7.10] and 0.625 obtained by ISAACSON and LUBENSKY [7.8] using a Flory approach.

Table 7.3. The exponent ν and the threshold parameter K_c for d = 2 SAWs (linear polymers) and lattice animals (branched polymers). The upper number is the estimated ν, obtained from the direct position-space renormalization group (PSRG) described in Sects.7.3.2,3. The lower number, in parenthesis, is the prediction for the threshold parameter K_c. Adapted from [7.97]

	Direct PSRG	Other methods
Linear	0.756 ± 0.004[a] (0.3791 ± 0.0001)	0.74 - 0.75[c] (0.3790[d])
Branched (with loops)	0.627[b] (0.254)	0.61 - 0.65[e] (0.192[f])

[a]Extrapolation of b = 2 - 150 results (cf. Table 7.2) [7.98]; [b]a cell-to-cell transformation with b_1 = 3, b_2 = 2 [7.97]; [c]FLORY [7.2] (ν = 0.75); KREMER et al. [7.107] (ν = 0.74 ± 0.01); DERRIDA [7.99] (ν = 0.7503 ± 0.0002); [d]SYKES et al. [7.47]; [e]ISAACSON and LUBENSKY [7.8] (ν = 0.625); PARISI and SOURLAS [7.10] (ν = 0.61); GOULD and HOLL [7.108] (ν = 0.65 ± 0.02); DERRIDA and de SEZE [7.109] (ν = 0.6408 ± 0.0003); [f]SYKES and GLEN [7.48]

The discrepancy becomes smaller when one calculates K'(K) for b = 3: K^* increases to 0.261, and ν increases to 0.593. A cell-to-cell transformation based on the b = 2 and b = 3 calculations gives K^* = 0.254 and ν = 0.627 (cf. Table 7.3).

Excluding loop formation leads to a recursion relation analogous to (7.31). The corresponding exponent ν differs so little from the value obtained from the full recursion relation [7.97], that it lends support to the prediction from field theoretical calculations that branched polymers with and without loops belong to the same universality class [7.11].

b) *Extensions*

1) Crossover Between SAWs and Lattice Animals. Given the parallels between the treatment of the SAW and lattice animal problems, it is tempting to incorporate both problems in a single two-parameter model. Let K be the monomer fugacity as before, and let f represent the probability that a randomly chosen site on a polymer is a polyfunctional unit (i.e., that this site is free to have more than two bonds "branching out" from it). One obtains *coupled* recursion relations [7.97] of the form,

$$K' = K'(K,f) \quad , \tag{7.32a}$$

$$f' = f'(K,f) \quad . \tag{7.32b}$$

This two-parameter position-space renormalization group transformation may be solved numerically for the fixed points, critical surface, and critical exponents [7.97]. The global flow diagram is sketched in Fig.7.14a for the special case of no loops. Note that there is a critical fixed point with f = 0 that is clearly identifiable

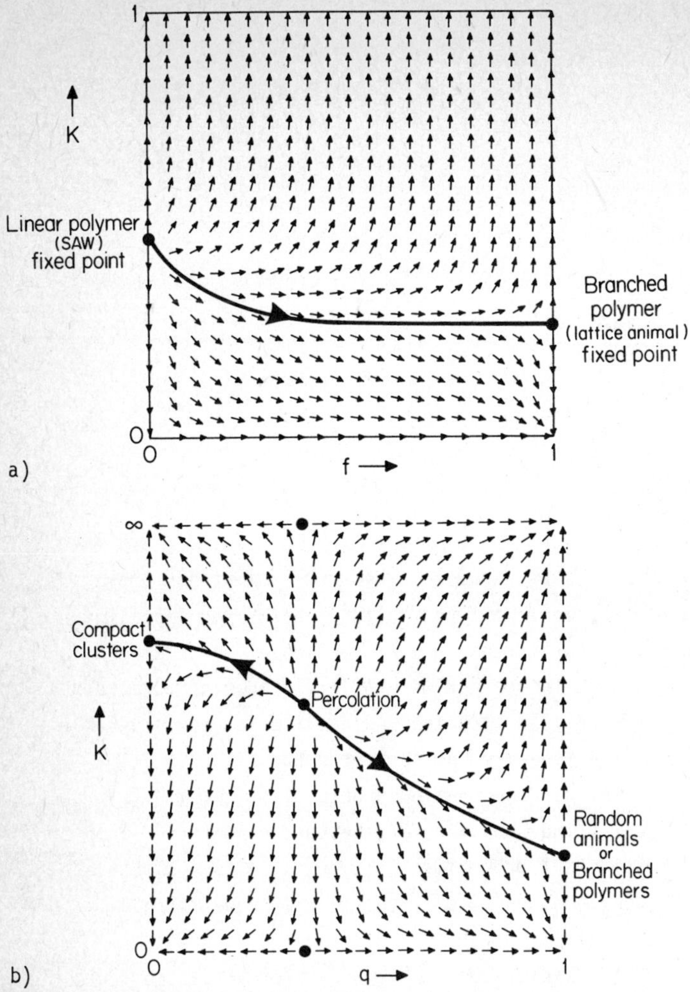

Fig.7.14. (a) Flow diagram showing crossover from SAWs to lattice animals as f, the probability of a polyfunctional unit, is varied. It is seen that SAWs and lattice animals belong to different universality classes. When $f > 0$, the flow is to the lattice animal fixed point. (b) Flow diagram for the crossover between lattice animals, percolation, and compact clusters, the 3 limits of the generalized animal problem cf. (7.33) . At $q = 0$, only clusters with the smallest possible perimeters contribute (compact clusters), while at $q = 1$ all lattice animals contribute equally (the "random animal" limit). Finally, along the line $q = 1 - K$, one obtains the percolation problem. The flow diagram suggests that these 3 cases belong to 3 separate universality classes. The percolation point is a multicritical point from which the flow is to the random animal fixed point for $q > q_c$, and to the compact cluster point for $q < q_c$. Adapted from [7.15,97]

as the SAW fixed point since no branching is allowed, and a second critical fixed point with $f = 1$, that corresponds to a structure that is fully branched. Additionally, for any amount of branching, no matter how small, the flow is toward the latter fixed point. Physically, this suggests that when even a few cross-linking units are present, polymeric systems obey the statistics of fully branched polymers. However, SAWs and lattice animals belong to two different universality classes. By following the flow trajectory from the SAW fixed point toward the lattice animal fixed point, one obtains the critical surface—shown as a solid line in Fig.7.14a.

This result agrees with a "blob" picture of randomly branched polymers [7.9]. The introduction of cross-linking units imposes a length scale, R_{blob}, defined by the mean separation between the cross links. One can thereby consider "effective monomers" of diameter, R_{blob} undergoing polyfunctional condensation. Thus for distance scales larger than R_{blob}, one expects to observe the asymptotic behavior of branched polymers.

2) Crossover Between Lattice Animals and Percolation. Consider the generalized generating function [7.15]

$$G(K,q) = \sum_{config} c(L,t)\, K^L q^t \quad , \tag{7.33}$$

where q may be considered an arbitrary parameter. Here the summation is over all configurations of the desired topological type, having L bonds and t perimeter bonds. The topological type is specified by the particular configurations included in c(L,t). Depending on the weight q assigned to the perimeters, branched configurations with different statistics will occur [7.15]. Both the SAW and lattice animal problems can be obtained in the limit that the fugacity for perimeter bonds, q, is unity. Furthermore, if K is identified with the bond occupation probability, and q = 1 - K, then G(K,q) is identical to the generating function for the bond percolation problem.[10] Thus we can calculate the statistical properties [cf., e.g., (7.1)] of percolation clusters and other "general nonrandom animals" (i.e., $q \neq 1$) equally readily using the direct position-space renormalization group.

To illustrate this in terms of a site description, a renormalized cell now plays the role of either an interior (occupied) site or a perimeter (empty) site; here we formulate the discussion in terms of site animals, but the same procedure is equally applicable for bond animals. Each *site* in an animal is weighted by a fugacity K, and an occupied *cell* has a fugacity K'. In addition, perimeter sites receive a weight q, and a cell "occupied" by a perimeter site receives weight q'. To write the recursion relations for K' and q' we must define a rule for when a cell is occupied. If a cell is occupied when a cluster spans the cell in both horizontal and vertical directions (this rule is called R_2 in [7.44]), then the recursion relation for K is

$$K' = K^4 + 4K^3 q \quad . \tag{7.34a}$$

10 In this case, limiting c(L,t) to SAWs would result in a "restricted valence" percolation problem ([7.110] and references therein) in which lattice animals with more than two occupied bonds incident upon a given site, or with ring closures, are forbidden.

To renormalize q, we treat it as a probability that a site is empty (as in percolation) and write

$$q' = q^4 + 4q^3(1 - q) + 6q^2(1 - q)^2 \quad . \tag{7.34b}$$

Note that q is an independent variable in (7.34), i.e., $q \neq 1 - K$ in contrast to the percolation recursion relation. However, if $q = 1 - K$, the two equations are consistent, and they reduce to the recursion relations for percolation under rule R_2.

The coupled recursion relations (7.34a,b) constitute a two-parameter position-space renormalization group transformation for the "generalized animal" problem. Nine fixed points are found, as indicated schematically in the flow diagram of Fig.7.14b. The three *critical* fixed points are indicated by heavy dots. The most unstable fixed point is the "percolation fixed point" where $q = 1 - K$. If the system is initially on the critical surface near the percolation fixed point, it flows under successive renormalizations to either the "random lattice-animal fixed point" at $q = 1$ or to the "compact cluster fixed point" at $q = 0$. The critical behavior below p_c is governed by the random animal fixed point, while the critical behavior above p_c is controlled by the compact cluster fixed point. This procedure has been used [7.15] to confirm earlier conjectures of various workers [7.30] that the critical exponent ν (called ρ in much of the literature) governing the dependence of the mean diameter on the degree of polymerization should be a discontinuous function of p, taking on the random animal value for all $p < p_c$, the percolation value at $p = p_c$, and the compact cluster value for $p > p_c$.

The corresponding numerical values of the exponent ν may be calculated by computing the matrix of derivatives

$$T = \begin{pmatrix} \partial K'/\partial K & \partial K'/\partial q \\ \partial q'/\partial K & \partial q'/\partial q \end{pmatrix}_{K = K^*,\ q = q^*} \quad . \tag{7.35}$$

Since $\partial q'/\partial K = 0$, the eigenvalues of this matrix are simply

$$\lambda_K = \partial K'/\partial K|_{K^*,q^*} = 4K^3 + 12K^2q|_{K^*,q^*} \tag{7.36a}$$

and

$$\lambda_q = \partial q'/\partial q|_{K^*,q^*} = 12q^3 - 24q^2 + 12q|_{K^*,q^*} \quad . \tag{7.36b}$$

The exponent ν is given by $\ln b/\ln\lambda_K$, while ν_p, the percolation connectedness length exponent, would be obtained from this *same* expression if $q = 1 - K$ in (7.34a). This, however, is equivalent to (7.34b). Thus $\nu_p = \ln b/\ln\lambda_q$. The numerical values of the exponents are given in Table 7.4. In particular, we note that the result $\nu = 0.52 \pm 0.02$ at p_c has recently been obtained using the above approach by extrapolation of results for $b \leq 5$ [7.111].

Table 7.4. Results for the generalized animal problem, based on direct position-space renormalization group calculations on the square lattice using the smallest cell (b = 2). Best estimates for the exponents ν and ν_p are given in square brackets (see also Fig.7.14b)

	(K^*,q^*)	ν	ν_{perc}
Branched polymer (Random animal)	(0.47,1)	0.61 [0.61 - 0.65][a]	-
Percolation	(0.77,0.23)	0.56 [0.52 - 0.53][b]	1.39 [4/3][c]
Compact	(1,0)	1/2 [1/2][d]	-

[a]ISAACSON and LUBENSKY [7.8]; PARISI and SOURLAS [7.10]; FAMILY [7.97]; [b]GOULD and HOLL [7.108]; FAMILY and REYNOLDS [7.111]; [c]BLACK and EMERY [7.69]; NIENHUIS [7.70]; [d]Exact

7.4 Other Approaches

For reasons of clarity, we have presented one main avenue of approach to position-space renormalization group in percolation and polymers. It has the advantage that the same approach works for many different lattice statistical problems. Other approaches which show promise have also been recently initiated. We shall devote this concluding section to a brief discussion of some of these.

7.4.1 Percolation

For percolation, YOUNG and STINCHCOMBE [7.62] initiated the use of a decimation transformation as mentioned in Sect.7.3.1. This has the advantage of simplicity, but the drawback that thus far it has been largely a "single-shot" technique with no systematic method of gauging the magnitude of the errors introduced by the method.

Phenomenological renormalization [7.99,109,112,113] holds promise for providing extremely accurate estimates of critical parameters because, like the cell-to-cell approach, the convergence seems to be linear in 1/b, rather than in 1/ln b [cf. (7.16b)]. Moreover, fairly large b can be treated; currently, extrapolations are based on up to 10 exact data points. While the phenomenological renormalization approach gives impressive numerical results, it cannot be easily adapted to a multi-parameter rescaling. In contrast, the cell renormalization discussed above can be easily generalized to more parameters in order to treat crossover phenomena and global phase diagrams (see, e.g., Sect.7.3.3). The phenomenological method has been discussed in Chap.3 and so will not be further described here.

7.4.2 Self-Avoiding Walks

It was remarked above that the initial position-space renormalization group calculations for percolation utilized the Q-state Potts Hamiltonian and subsequently took the $Q \to 1$ limit to recover percolation properties. Similarly, the first position-space renormalization group calculations for SAWs utilized a discrete version of the "n-vector model" and subsequently took the $n \to 0$ limit ([7.114]; for a generalization of this model designed to model branched polymers see [7.115]).

Possible "direct" position-space renormalization group calculations that do *not* require using a Hamiltonian have been suggested by several authors [7.1,95-99]. The cell approach [7.96-98] has been described above because it is conceptually analogous to the approach used for both percolation and lattice animals. Now we briefly describe the other "direct" position-space renormalization group treatments of SAWs.

SHAPIRO [7.95] was the first to apply a decimation procedure directly on the SAW generating function. This was a very useful conceptual advance. NAPIÓRKOWSKI et al. [7.100] considered a cell position-space renormalization group using a second-order cumulant-like approximation requiring five parameters. They obtained $\nu = 0.712$ and 0.769 on the square and triangular lattices respectively. MALAKIS [7.116], following NAPIÓRKOWSKI et al., used a decimation technique with several parameters.

De GENNES [7.1] has suggested a variation of decimation that he terms "decimation along the chain". This general approach has been implemented and extended with good results. BAUMGÄRTNER [7.117] has used a Monte Carlo renormalization group version of it to obtain $\nu = 0.586 \pm 0.004$ for a *continuum* model of a $d = 3$ polymer. However, more extensive work based on this method [7.107] has given error bars on ν which include the Flory value $\nu = 3/5$. Thus in three dimensions, it appears that ν may be less than 0.6, although the issue does not seem completely resolved. For $d = 2$ KREMER et al. [7.107] found $\nu = 0.74 \pm 0.01$ based on an extensive study using this Monte Carlo renormalization method.

The highest claimed accuracy for $d = 2$ is due to DERRIDA [7.99], who has applied extrapolation ideas to a phenomenological renormalization procedure. On the square lattice with periodic boundary conditions, he obtained $\nu = 0.7503 + 0.0002$. However, his results have the same non-monotonicities as the large-cell SAW approach described above, in the case of cells with free boundary conditions. We raise here the same question as we raised previously: Just because the results with a particular weight function appear monotonic, are they therefore correct?

7.4.3 Lattice Animals

Very recently, the phenomenological renormalization approach has been extended to *site* lattice animals on a square lattice [7.109]. The extrapolated estimate $\nu = 0.6408 \pm 0.0003$ agrees well with estimates obtained by other methods. There is no position-space renormalization group work on *bond* animals other than that discussed in Sect.7.3.3.

7.5 Concluding Remarks and Outlook

In this chapter, we have focused on the application of a relatively simple approach — use of a "connectivity" weight function — to the description of critical phenomena occurring in three mathematical models of polymer systems. It is perhaps appropriate to conclude by emphasizing the generality of the connectivity approach.

Not surprisingly, this approach can be used to treat a wide range of problems for which the basic physics is "connectivity" — ranging from simple models such as those described in this chapter, to more complex models like those used to describe physical phenomena such as vulcanization [7.118,119], directed percolation [7.120-127], anisotropic percolation [7.128-131], second-neighbor percolation [7.129,132,133], site-bond percolation [7.31-33], correlated percolation [7.134,135], time-dependent percolation [7.108], and the related problem of "growing lattice animals" [7.136].

Perhaps more surprisingly, the connectivity approach can be extended to treat lattice statistical models for which connectivity does not at first sight appear to be the appropriate physical concept. For example, the essential physics underlying the thermal behavior of an Ising model is not obviously connectivity, yet we know that there are striking analogies between the Ising model behavior near T_c and percolation phenomena near p_c (see Fig.7.15). Recently, TSALLIS and his collaborators [7.138,139] have proposed an ingenious approach to the Ising and q-state Potts models which focuses upon the "transmission" of correlation through the bonds. They use the self-dual bond cells illustrated in Fig.7.12 rather than the more conventional Kadanoff construction using site cells. In this fashion they have constructed an Ising-model renormalization group, using connectivity weight functions, that correctly predicts the exact critical point for all finite-size cells.

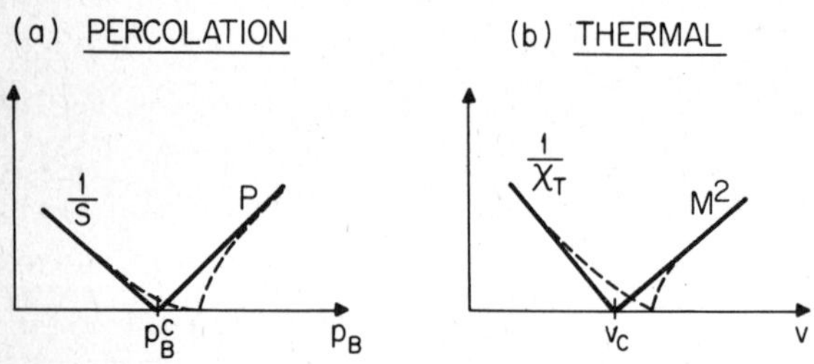

Bond probability p_B^c = percolation pt.	Coupling constant $v_c = \tanh(J/kT_c)$
Bond cluster	Droplet
P(p) S(p)	M χ_T
< $N_{clusters}$ >	G(H,T)
Fisher-Essam $d_c = 6$	Husimi-Temperley $d_c = (2\beta+\gamma)/\nu = 4$

Fig.7.15a,b. Illustration of the analogies between (a) bond percolation, and (b) an ordinary thermal phase transition (e.g., an Ising or lattice-gas model). Adapted from [7.137]. As discussed in Sect.7.5, the same connectivity approach used in percolation can be extended to the "thermal transmissivity" of the magnetic system

While we have not emphasized the generality of the connectivity approach, we have tried to emphasize the extent to which the "one-cell, one-parameter renormalization group" can attain high accuracy by considering successively larger cells. Thus while other position-space renormalization group approaches include more parameters in an effort to obtain better accuracy, this approach focuses upon using larger cells. The advantage of our approach is that there is a well-defined procedure for extrapolating a sequence of approximate results. The outcomes of such extrapolations give numerical agreement with exact information where available.

There is one circumstance where it is desirable—and feasible—to include additional parameters. This occurs when one wishes to study crossover between universality classes. Such studies were illustrated above in Sect.7.3.3 when we discussed the crossover between linear and branched polymers, and the crossover between branched polymers and percolation. One may thereby determine if an additional parameter is relevant or irrelevant. Thus, e.g., site percolation and bond percolation can be clearly demonstrated to belong to the same universality class [7.32,33]. Four-coordinated correlated percolation [7.140] and random percolation can be similarly shown to be in the same universality class [7.135]. In these cases, the additional parameter which was introduced was an irrelevant variable. On the other hand, e.g., directed ("diode") percolation and random percolation can clearly be seen to belong to different universality classes [7.122], and the additional variable is relevant.

Acknowledgements. Some of the work described here was carried out in collaboration with other members of our research group, including Antonio Coniglio, Agustin E. Gonzalez, Harvey Gould, William Klein, Hisao Nakanishi, and Gerald Shlifer. We are particularly indebted to sustained interactions with our distinguished visitors, especially Professors Kurt Binder, Pierre-Gilles de Gennes, Dietrich Stauffer, and Constantino Tsallis, who have generously shared with us their interest and enthusiasm. Finally, we wish to thank Bernard Derrida and Robin Stinchcombe for continuing to send us copies of their groups' work prior to publication. Also, we wish to apologize to these and perhaps other authors for the fact that space—and considerations of pedagogy—do not permit a fuller exposition of their important contributions to this field.

References

7.1 P.G. de Gennes: *Scaling Concepts in Polymer Physics* (Cornell Univ. Press, Ithaca 1979); Riv. Nuovo Cimento *7*, 363 (1972)

7.2 P.J. Flory: *Principles of Polymer Chemistry* (Cornell Univ. Press, Ithaca 1953)

7.3 M.E. Fisher: J. Phys. Soc. Jpn (Suppl.) *26*, 44 (1969)

7.4 J. des Cloiseaux: J. Phys. Paris *37*, 431 (1976)

7.5 P.G. de Gennes: Phys. Lett. *38*A, 339 (1972)

7.6 H.E. Stanley: Phys. Rev. Lett. *20*, 489 (1968)

7.7 M. Daoud, J.P. Cotton, B. Farnoux, G. Jannink, G. Sarma, H. Benoit, R. Duplessix, C. Picot, P.G. de Gennes: Macromolecules *8*, 804 (1975)

7.8 J. Isaacson, T.C. Lubensky: J. Phys. Paris Lett. *41*, L469 (1980)
7.9 M. Daoud, J.F. Joanny: J. Phys. Paris *42*, 1359 (1981)
7.10 G. Parisi, N. Sourlas: Phys. Rev. Lett. *46*, 871 (1981)
7.11 T.C. Lubensky, J. Isaacson: Phys. Rev. A*20*, 2130 (1979)
7.12 P.G. de Gennes: C.R. Acad. Sci. *291*, 17 (1980)
7.13 A.B. Harris, T.C. Lubensky: Phys. Rev. B*24*, 2656 (1981)
7.14 F.Y. Wu: Rev. Mod. Phys. *54*, 235 (1982)
7.15 F. Family, A. Coniglio: J. Phys. A*13*, L403 (1980)
7.16 H.E. Stanley: In *Festschrift for Laszlo Tisza*, ed. by A. Shimony, H. Feshbach (MIT Press, Cambridge 1982)
7.17 A.B. Harris, T.C. Lubensky, W.K. Holcomb, C. Dasgupta: Phys. Rev. Lett. *35*, 327 (1975)
7.18 R.G.Priest, T.C. Lubensky: Phys. Rev. B*13*, 4159 (1976)
7.19 D.J. Amit: J. Phys. A*9*, 1441 (1976)
7.20 M. Gordon, S.B. Ross-Murphy: Pure Appl. Chem. *43*, 1 (1975)
7.21 T. Vicsek, J. Kertész: J. Phys. A*14*, L31 (1981)
7.22 E.T. Gawlinski, H.E. Stanley: J. Phys. A*14*, L291 (1981);
S.W. Haan, R. Zwanzig: J. Phys. A*10*, 1547 (1977)
7.23 P.J. Flory: J. Am. Chem. Soc. *63*, 3083 (1941)
7.24 W.H. Stockmayer: J. Chem. Phys. *12*, 125 (1944)
7.25 M.E. Fisher, J.W. Essam: J. Math. Phys. *2*, 609 (1961)
7.26 H.L. Frisch, J.M. Hammersley: SIAM J. Appl. Math. *11*, 894 (1963)
7.27 P.G. de Gennes: J. Phys. Paris *36*, 1049 (1975)
7.28 D. Stauffer: J. Chem. Soc. Faraday Trans. II *72*, 1354 (1976)
7.29 D. Stauffer, A. Coniglio, M. Adam: Adv. Ply. Sci. *44*, 103 (1982)
7.30 D. Stauffer: Phys. Rep. *54*, 1 (1979)
7.31 P. Agrawal, S. Redner, P.J. Reynolds, H.E. Stanley: J. Phys. A*12*, 2073 (1979)
7.32 H. Nakanishi, P.J. Reynolds: Phys. Lett. *71*A, 252 (1979)
7.33 B. Shapiro: J. Phys. C*12*, 3185 (1979)
7.34 A. Coniglio, H.E. Stanley, W. Klein: Phys. Rev. Lett. *42*, 518 (1979)
7.35 A. Coniglio, H.E. Stanley, W. Klein: Phys. Rev. B*25*, 6805 (1982)
7.36 T. Tanaka, G. Swislow, A. Ohmine: Phys. Rev. Lett. *42*, 1556 (1979)
7.37 A.E. Gonzales, M. Daoud: J. Phys. A*14*, 2441 (1981)
7.38 A.E. Gonzales, S. Muto: J. Chem. Phys. *73*, 4668 (1980)
7.39 M. Schmidt, W. Burchard: Macromolecules *14*, 370 (1981)
7.40 C. Domb: J. Phys. A*9*, 283 (1976)
7.41a H.E. Stanley: J. Phys. A*10*, L211 (1977);
7.41b A. Coniglio: Phys. Rev. Lett. *46*, 250 (1981); Preprint (1981)
7.41c R. Pike, H.E. Stanley: J. Phys. A*14*, L169 (1981)
7.42 P.J. Reynolds, W. Klein, H.E. Stanley: J. Phys. C*10*, L167 (1977)
7.43 P.J. Reynolds, H.E. Stanley, W. Klein: J. Phys. A*11*, L199 (1978)
7.44 P.J. Reynolds, H.E. Stanley, W. Klein: Phys. Rev. B*21*, 1223 (1980)
7.45 C. Domb: Adv. Chem. Phys. *15*, 229 (1969)
7.46 D.S. McKenzie: Phys. Rep. *27*C, 35 (1976)
7.47 M.F. Sykes, A.J. Guttman, M.G. Watts, P.D. Roberts: J. Phys. A*5*, 653 (1972)
7.48 M.F. Sykes, M. Glen: J. Phys. A*9*, 87 (1976)
7.49 D.S. Gaunt: J. Phys. A*13*, L97 (1980)
7.50 J. Hoshen, D. Stauffer, G.H. Bishop, R.J. Harrison, G.B. Quinn: J. Phys. A*12*, 1285 (1979)
7.51 H. Nakanishi, H.E. Stanley: J. Phys. A*11*, L189 (1978); Phys. Rev. B*22*, 2466 (1980); J. Phys. A*14*, 693 (1981)
7.52 J.W. Essam: Rep. Prog. Phys. *43*, 833 (1980)
7.53 P.J. Reynolds, H.E. Stanley, W. Klein: J. Phys. A*10*, L203 (1977)
7.54 D.S. Gaunt, H. Ruskin: J. Phys. A*11*, 1369 (1978)
7.55 D.S. Gaunt, M.F. Sykes, H. Ruskin: J. Phys. A*9*, 1899 (1976)
7.56 T.D. Lee, C.N. Yang: Phys. Rev. *87*, 410 (1952)
7.57 R.B. Potts: Proc. Cambridge Philos. Soc. *48*, 106 (1952)
7.58 P. Kasteleyn, C.M. Fortuin: J. Phys. Soc. Jpn (Suppl.) *26*, 11 (1969)
7.59 D.J. Amit, D.J. Wallace, R.K.P. Zia: Phys. Rev. B*15*, 4657 (1977)
7.60 C. Dasgupta: Phys. Rev. B*14*, 1221 (1976)

7.61 T.S. Burkhardt, B.W. Southern: J. Phys. *A11*, L253 (1975)
7.62 A.P. Young, R.B. Stinchcombe: J. Phys. *C8*, L535 (1975); *9*, L643 (1976)
7.63 L.G. Marland, R.P. Stinchcombe: J. Phys. *C10*, 2223 (1977)
7.64 L.G. Marland: J. Phys. *C11*, L617 (1978)
7.65 S. Kirkpatrick: Phys. Rev. *B15*, 1533 (1977)
7.66 A.A. Migdal: Sov. Phys. JETP *42*, 413; 743 (1975)
L.P. Kadanoff: Ann. Phys. (N.Y.) *100*, 359 (1976)
7.67 Z. Djordjevic, H.E. Stanley, A. Margolina: J. Phys. *A15* (1982)
M.F. Sykes, D.S. Gaunt, M. Glen: J. Phys. *A9*, 97 (1976)
7.68 M.P.M. den Nijs: J. Phys. *A12*, 1857 (1979)
7.69 J.L. Black, V.J. Emery: Phys. Rev. *B23*, 429 (1981)
7.70 B. Nienhuis: J. Phys. *A15*, 199 (1982)
7.71 Z. Friedman, J. Felsteiner: Phys. Rev. *B15*, 5317 (1977)
7.72 M.E. Fisher: In *Critical Phenomena*, ed. by M.S. Green (Academic, New York 1971);
A. Sur, J.L. Lebowitz, J. Marro, M.L. Kalos, S. Kirkpatrick: J. Stat. Phys. *15*, 145 (1976)
7.73 E. Brézin: J. Phys. Paris *43*, 15 (1982)
7.74 P.D. Eschbach, D. Stauffer, H.J. Herrmann: Phys. Rev. *B23*, 422 (1981)
7.75 J. Hoshen, R. Kopelman: Phys. Rev. *B14*, 3438 (1976)
7.76 R.B. Pearson: Phys. Rev. *B22*, 2579 (1980)
7.77 B. Nienhuis, E.K. Riedel, M. Schick: J. Phys. *A13*, L189 (1980)
7.78 J. Bernasconi: Phys. Rev. *B18*, 2185 (1978)
7.79 S. Kirkpatrick: In *Ill-Condensed Matter*, ed. by R. Balian, R. Maynard, G. Toulouse (North Holland, Amsterdam 1979)
7.80 C.J. Lobb, K.R. Karasek: J. Phys. *C13*, L245 (1980);
C.J. Lobb, K.R. Karase: Phys. Rev. *B25*, 492 (1982)
7.81 A.C.N. de Magalhães, C. Tsallis, G. Schwachheim: J. Phys. *C13*, 321 (1980)
7.82 D.W. Heermann, D. Stauffer: Z. Phys. *B40*, 113 (1981); Preprint (1981)
7.83 Y. Yuge: J. Phys. *A12*, 2509 (1979); Phys. Rev. *B18*, 1514 (1978)
7.84 A.G. Dunn, J.W. Essam, D.S. Ritchie: J. Phys. *C8*, 4219 (1975)
7.85 R. Rosman, B. Shapiro: Phys. Rev. *B16*, 5117 (1977)
7.86 R.B. Stinchcombe, B.P. Watson: J. Phys. *C9*, 3221 (1976)
7.87 C.J. Lobb, D.J. Frank: J. Phys. *C12*, L827 (1979);
AIP Conf. Proc. *58*, 308 (1980);
C.J. Lobb, D.J. Frank, M. Tinkham: Phys. Rev. *B23*, 2262 (1981)
7.88 H. Kunz, B. Payandeh: Phys. Rev. *B20*, 1285 (1979)
7.89 R. Fisch, A.B. Harris: Phys. Rev. *B18*, 416 (1978)
7.90 G. Shlifer, W. Klein, P.J. Reynolds, H.E. Stanley: J. Phys. *A12*, L169 (1979)
7.91 S. Kirkpatrick: AIP Conf. Proc. *40*, 99 (1978)
7.92 R. Pike, H.E. Stanley: J. Phys. *A14*, L169 (1981)
7.93 W. Kinzel: Z. Phys. *B34*, 79 (1979)
7.94 H.E. Stanley, R.J. Birgeneau, P.J. Reynolds, J.F. Nicoll: J. Phys. *C9*, L553 (1976)
7.95 B. Shapiro: J. Phys. *C11*, 2829 (1978)
7.96 S.L.A. de Queiroz, C.M. Chaves: Z. Phys. *B40*, 99 (1980)
7.97 F. Family: J. Phys. *A13*, L325 (1980)
7.98 S. Redner, P.J. Reynolds: J. Phys. *A14*, 2679 (1981); *14*, L55 (1981)
7.99 B. Derrida: J. Phys. *A14*, L5 (1981)
7.100 M. Napiórkowski, E.M. Hauge, P.C. Hemmer: Phys. Lett. *72*A, 193 (1979)
7.101 F. Family: J. Phys. Paris *42*, 189 (1981)
7.102 M.G. Watts: J. Phys. *A8*, 61 (1975)
7.103 J.C. le Guillou, J. Zinn-Justin: Phys. Rev. *B21*, 3976 (1980)
7.104 J.P. Cotton: J. Phys. Lett. Paris *41*, L231 (1980)
7.105 B. Shapiro: J. Phys. *C11*, L863 (1978)
7.106 B. Shapiro: J. Phys. *C13*, 3387 (1980)
7.107 K. Kremer, A. Baumgärtner, K. Binder: Z. Phys. *B40*, 331 (1981)
7.108 H. Gould, K. Holl: J. Phys. *A14*, L443 (1981)
7.109 B. Derrida, L. de Seze: J. Phys. Paris *43*, 475 (1982)
7.110 J. Kertész, B.K. Chakrabarti, J.A.M.S. Duarte: J. Phys. *A15*, L13 (1982);
S.G. Whittington, K.M. Middlemiss, D.S. Gaunt: J. Phys. *A14*, 2415 (1981);

S.G. Whittington, K.M. Middlemiss, G.M. Torrie, D.S. Gaunt: J. Phys. *A13*, 3707 (1980);
R. Cherry, C. Domb: J. Phys. *A13*, 1325 (1980);
D.S. Gaunt, J.L. Martin: G. Ord, G.M. Torrie, S.G. Whittington: J. Phys. *A13*, 1791 (1980)
7.111 F. Family, P.J. Reynolds: Z. Phys. *B45*, 123 (1981)
7.112 M.P. Nightingale: Proc. K. Ned. Ak. Wet. *B82*, 235 (1979)
7.113 B. Derrida, J. Vannimenus: J. Phys. Paris Lett. *41*, L473 (1980);
H.W.J. Blöte, M.P. Nightingale, B. Derrida: J. Phys. *A14*, L45 (1981)
7.114 H.J. Hilhorst: Phys. Lett. *56*A, 153 (1976); Phys. Rev. *B16*, 1253 (1977)
7.115 T.C. Lubensky, C. Dasgupta, C.M. Chaves: J. Phys. *A11*, 2219 (1978)
7.116 A. Malakis: Physica *104*A, 427 (1980)
7.117 A. Baumgärtner: J. Phys. *A13*, L39 (1980)
7.118 A. Coniglio, M. Daoud: J. Phys. *A12*, L259 (1979)
7.119 M. Daoud, A. Coniglio: J. Phys. *A14*, L108 (1981)
7.120 J. Kertész, T. Vicsek: J. Phys. *C13*, L343 (1980)
7.121 P.J. Reynolds: Preprint (1981)
7.122 S. Redner: J. Phys. *A14*, L349 (1981);
S. Redner, A.C. Brown: J. Phys. *A14*, L285 (1981)
7.123 W. Kinzel, J. Yeomans: J. Phys. *A14*, L163 (1981)
7.124 W. Klein, W. Kinzel: J. Phys. *A14*, L405 (1981);
W. Klein: J. Phys. *A15* (1982)
7.125 S. Redner, Z.R. Yang: J. Phys. *A15*, 77 (1982)
7.126 S. Redner: Phys. Rev. *B25*, 3242 (1982)
7.127 S. Redner: Phys. Rev. *B25*, 5646 (1982)
7.128 H. Nakanishi, P.J. Reynolds, S. Redner: J. Phys. *A14*, 855 (1981);
H. Ikeda: Prog. Theor. Phys. *61*, 842 (1979)
7.129 A.C.N. de Magalhães, C. Tsallis, G. Schwachheim: J. Phys. *C14*, 1393 (1981)
7.130 C.M. Chaves, P.M. Oliveira, S.L.A. de Queiroz: Prog. Theor. Phys. *62*, 1550 (1979)
7.131 P. Murilo Oliveira: Phys. Rev. *B25* (1982)
7.132 P. Murilo Oliveira: Phys. Rev. Lett. *47*, 1423 (1981)
7.133 M. Napiórkowski, P.C. Hemmer: Phys. Lett. *76*A, 359 (1980)
7.134 M. Napiórkowski, P.C. Hemmer: Phys. Lett. *75*A, 153 (1980)
7.135 A. Gonzales, P.J. Reynolds: Phys. Lett. *80*A, 357 (1980)
7.136 H. Gould, F. Family, H.E. Stanley: Preprint (1982);
H.P. Peters, D. Stauffer, H.P. Hölters, D. Loewenich: Z. Phys. *B34*, 339 (1979)
7.137 H.E. Stanley, J. Teixeira: J. Chem. Phys. *73*, 3404 (1980)
7.138 C. Tsallis, G. Schwachheim: J. Phys. *C12*, 9 (1979);
C. Tsallis, S.V.F. Levy: J. Phys. *C13*, 465 (1980);
S.V.F. Levy, C. Tsallis, E.M.F. Curado: Phys. Rev. *B21*, 2991 (1980)
7.139 C. Tsallis: J. Phys. *C13*, L85 (1981);
F.C. Alcaraz, C. Tsallis: J. Phys. *A15*, 587 (1982)
7.140 H.E. Stanley: J. Phys. *A12*, L329 (1979)

Subject Index